PRAISE FOR

SICKER, FATTER, POORER

"In the spirit of *Silent Spring,* Dr. Trasande beautifully couples science and action, laying out resounding evidence of hormone disruption and charging all of us to do better—for our children and our children's children."

—Mona Hanna-Attisha, M.D., M.P.H., author of
*What the Eyes Don't See: A Story of Crisis, Resistance,
and Hope in an American City*

"Dr. Leonardo Trasande is the new Oliver Sacks. If you want to understand how chemicals are hacking hormones and fueling epidemics like autism and ADHD, then you have found an explainer-extraordinaire with a wonderful bedside manner."

—Pete Myers, co-author of *Our Stolen Future*

"A lively, readable guide to the risks and costs of our social addiction to endocrine-disrupting compounds. It's an important but often bewildering area of science that only someone with Dr. Trasande's deep knowledge and true passion for the subject could have untangled."

—Dan Fagin, author of the Pulitzer Prize–winning
Toms River: A Story of Science and Salvation

"An incredible tool for business leaders and wellness warriors alike. Dr. Trasande weaves together the latest knowledge, empowering you to incorporate small changes that will make a huge difference in your health. His prescription for business is

also timely: offering products with safer ingredients and materials is good for the bottom line too."
— Christopher Gavigan, author of *Healthy Child Healthy World*, and co-founder of The Honest Company

"Escalating obesity, diabetes, learning disorders, autism, infertility, and food allergies are often presented as a medical mystery, but scientific evidence shows that many of them result from endocrine-disrupting chemicals in our food, our homes, our personal care products. In this urgent book, Dr. Trasande explains the science behind these disturbing trends, who is responsible, and what we can do about it."
— Naomi Oreskes, co-author of *Merchants of Doubt: How a Handful of Scientists Obscured the Truth on Issues from Tobacco Smoke to Global Warming*

SICKER
FATTER
POORER

THE URGENT THREAT OF

HORMONE-DISRUPTING CHEMICALS

TO OUR HEALTH AND FUTURE . . .

AND WHAT WE CAN DO ABOUT IT

LEONARDO TRASANDE, M.D., M.P.P.

MARINER BOOKS

HOUGHTON MIFFLIN HARCOURT

Boston   New York

The information contained in this book is intended to provide helpful and informative material on the subject addressed. It is not intended to serve as a replacement for professional medical advice. Any use of the information in this book is at the reader's discretion. The author and publisher disclaim any and all liability arising directly and indirectly from the use or application of any information contained in this book.

First Mariner Books edition 2021

For information about permission to reproduce selections from this book, write to trade.permissions@hmhco.com or to Permissions, Houghton Mifflin Harcourt Publishing Company, 3 Park Avenue, 19th Floor, New York, New York 10016.

hmhbooks.com

Library of Congress Cataloging-in-Publication Data
Names: Trasande, Leonardo, author.
Title: Sicker, fatter, poorer : the urgent threat of hormone-disrupting chemicals on our health and future . . . and what we can do about it / Leonardo Trasande, MD, MPP.
Description: Boston : Houghton Mifflin Harcourt, [2019] | Includes bibliographical references and index.
Identifiers: LCCN 2018032625 (print) | LCCN 2018053000 (ebook) | ISBN 9781328553805 (ebook) | ISBN 9781328553492 (hardcover) | ISBN 9780358410966 (paperback)
Subjects: LCSH: Endocrine disrupting chemicals — Health aspects.
Classification: LCC RC649 (ebook) | LCC RC649 .T73 2019 (print) | DDC 362.196/4 — dc23

LC record available at https://lccn.loc.gov/2018032625

Book design by Kelly Dubeau Smydra

Printed in the United States of America
DOC 10 9 8 7 6 5 4 3 2 1

In memory of Rachel Carson, Theo Colborn and Lou Guillette. Your legacies inspire us all to carry on your work to protect human health and the environment from endocrine-disrupting chemicals.

For my two sons, Camilo and Ramiro, whose future I think about each day.

CONTENTS

INTRODUCTION

Thousands of chemicals are negatively affecting our brains, bodies, and environment each and every day. Substances invisible to the naked eye are not only disrupting the most important hormones of our body and brain but also laying down multiple paths of disease that will impact our children and their children decades into the future. These are challenging and uncomfortable ideas to accept. You might want to throw up your hands when you hear that these chemicals are produced and distributed on a massive scale, are minimally regulated, and will continue to wreak havoc on our lives, generation after generation.

As mind-boggling as it is, this dire scenario is all too real.

I trust that you know someone who has attention-deficit/ hyperactivity disorder (ADHD) or whose child or grandchild has been diagnosed with autism. You may have wondered about the dramatic increase in obesity and diabetes across American society over the past couple of decades. Study after study has shown that these increases can be directly tied to chemicals in our food supply, environment, and household and personal care products. You may not yet be aware of the slow yet steady rise in fertility issues experienced by both women and men. You may not be tuned in to the documented drop in sperm count for men as young as in their 20s. You may not yet appreciate the long-term economic costs to society when young children are born with an already-diminished IQ. These, too, are some of the frightening and real outcomes of endocrine disruption.

One of the first major public disclosures that the widespread use of synthetic chemicals can do harm—not just provide benefits—occurred when Rachel Carson's *Silent Spring* was published in 1962. Though scientists had been questioning and investigating the deleterious effects of man-made or synthetic chemicals since World War I, the wider world was just beginning to wake up to the dangers of chemicals such as DDT, sprayed over farmland, wetlands, and in our neighborhoods, to our natural habitat. Half a century later, Carson's deeply felt exploration of the harms of pesticides is not only still relevant but also serves as a dramatic reminder that we have not fully resolved the issues that she raised more than five decades ago. Indeed, our environment and human health and survival are more endangered now than ever before.

Like many of us, I am deeply concerned about the degree to which chemicals are still infiltrating our world, our bodies, and our brains. I am as much an environmentalist as any sane person troubled by the state of the air we breathe, the water we drink, and the land from which we harvest our food. But I am first and foremost a pediatrician—and the father of two young children. My education and training began at Harvard College and then Harvard Medical School. During medical school interviews, I recall frequently being asked for my opinion about President Clinton's proposal for universal health care. Everyone seemed to have an opinion one way or another, but I wasn't sure I had the proper training to answer this important policy question.

I began to ask whether students were taught how health care policy is made and how doctors could shape the process. I soon realized how little room there was in the medical school curriculum to focus on the political and economic forces that shape medicine. Indeed, to this day, very little about health policy is included in the typical medical school curriculum, especially policy that is related to environmental health and the impact of chemicals. In order to gain a more informed perspective, I enrolled at the Harvard Kennedy School of Government, where I dived deep into health policy and its economics. This experience fundamentally transformed the way I thought about medical care. I realized that I had the opportunity to help many more people if I could bring my knowledge of medicine to the process of shaping decisions that could affect thousands if not millions of people.

Soon after I completed my pediatric residency at Boston Children's Hospital and Boston Medical Center, I was accepted into a fellowship program with then-Senator Hillary Clinton's office. I was asked to focus on two areas—children's health and environmental health—and was excited to work on child health policy. I cared about the environment but had not yet connected it to the health of children in a very deep way.

My time working for then-senator Clinton was another transformative experience. I came to more fully appreciate the enormous and growing impact of chemicals and other environmental factors on human health and the role of regulation in eliminating or reducing the most harmful exposures. I finished my fellowship with a new and energized career focus: to study the impact of environmental exposures on children and to document the benefits of prevention to society at large.

After additional training in environmental medicine, I embarked on a career at the intersection of environmental science, medicine, and policy. Currently, my research focuses on identifying preventable and environmental factors that contribute to obesity and other chronic conditions in children, as well as measuring and documenting the economic costs of failing to prevent environmental hazards. I now serve as an associate professor of pediatrics at the NYU School of Medicine, of health policy at the NYU Wagner School of Public Service and at the NYU College of Global Public Health. I serve on several national and international committees that study the impact of chemicals on children. I have been working in environmental health for close to 20 years.

For me, this book is an extension of my commitment to ensure that everyone has the opportunity to understand the long-term threat of synthetic chemicals and their relationship to endocrine disruption. I also hope to grow your appreciation that, though it's hard to measure the risks in the now, many chemicals have the power—and the tendency—to show up down the road, when it's often too late. In fact, that's why this book is both so timely and so pressing. The longer we wait to act, the longer we wait to truly take the power we do possess as citizens, to change our own habits and ultimately affect governmental policy, the more the danger grows. It can impact our own health and the long-term health of our children and grandchildren. Many health risks may not be crystal clear right now; however, the science itself is quite clear, pointing to illnesses, an increase in obesity, and shockingly the loss of IQ points that translates to a decrease in one's ability to earn money.

This may all sound dire and frightening. However, I want to assure you that there is hope in this message and in this book, especially in its concluding two chapters. There are actions you can take to protect yourself and your family, and there is a larger virtuous circle awaiting your participation.

MY STORY

In 2014, I was asked to organize the Endocrine Disrupting Chemical (EDC) Disease Burden Working Group, a group of nearly 30 scientists from eight countries that advises their policymakers about the costs of conditions that can be linked to chemicals that can disrupt hormones in our bodies. (Throughout the book I will refer to this group of internationally regarded scientists, medical doctors, and other experts in endocrine disruption.) Our findings revealed that the chemicals referred to here are responsible for additional disease-related costs in Europe of €163 billion (which equates to $209 billion) each year.[1,2] How many chemicals did my group study? Fewer than 5% of all known hormone-disrupting chemicals. And yet the literal billions of euros (translated into even more US dollars) is likely a grave understatement of the overall problem. Because we had to restrict the number of chemicals in our study, our team could only begin to imagine the harmful effects of the thousands of other chemicals we have not yet had a chance to examine. For the sake of streamlining our studies, we chose not to include many conditions and diseases that are indirectly implicated by chemical exposure, including prostate cancer, osteoporosis, breast cancer, and certain immune conditions—adult diseases triggered as early as childhood, sometimes even *in utero*, after exposure to chemicals. Even the most sophisticated studies aimed at understanding these chemicals have only begun to scratch the surface of a much, much bigger problem.

In this book, I have chosen to focus on a few groups of chemicals because it's important to me to stay as closely aligned with the best scientific evidence possible. Though many other groups of chemicals may pose similar dangers, the scientific research is strongest and most current for the groups of chemicals described here. So in this book you will learn about how certain groups of chemicals disrupt our hormonal systems and damage our health in irreparable ways. You will come to learn about these chemicals and where they hide—in your home, workplace, the foods that you eat, and in surprising areas of your everyday environment. You will also begin to gain an understanding of how these chemicals affect your and your family's brains and bodies and just how tricky they can be to trace. Unlike an infection, whose cause can be traced through microbiological analysis, chemicals can work almost invisibly. Routine clinical tests or exams may not detect exposures, or the traces of chemicals may disappear long before their effects emerge.

The book will describe

- how hormone-disrupting chemicals get into our bodies;
- how they can mimic our hormones and what happens when they do;
- how hormone disruption can contribute to a broad array of diseases, including brain disorders, metabolic disruptions such as type 2 diabetes and obesity, and reproductive illnesses;
- how the increased incidents of these diseases impact individuals and our society more broadly;
- how we can all limit our exposure to the chemicals of greatest concern while maintaining our urban, suburban, or rural lifestyles of choice;
- what the forces—political, economic, and policymaking— were that led to the epidemic we are facing today, how the regulatory framework is flawed, and how your consumer purchasing power holds promise in compensating for the absence of policy change, as it did with removing lead from paint, bisphenol A from baby bottles, Alar from apple juice, and many other examples.

You will come to know the products commonly used in your family's everyday lives that contain endocrine-disrupting chemicals (EDCs) and the simple, immediate steps you can take to lessen their danger. I've also included some case studies so that you can appreciate the toll these chemicals can take in real life. The case studies I describe are composites of multiple patients I've encountered throughout my career, with names and details intentionally modified to protect privacy and emphasize the key takeaways. Although many researchers step away from clinical care, to this day I still practice clinical pediatrics and work with patients and their families so that I can stay grounded in the real-life situations experienced by real people. As much as I am a policy wonk, staying close to children and their families informs the research I do and helps me better communicate the implications of my research to policymakers.

I have also incorporated the most important research studies so that you can appreciate why my colleagues in the field of endocrine disruption and I are so concerned and why you should be, too. There are prominent public figures who undermine the significance of these health issues and claim the science is wrong. There are some even within the medical community who attempt to minimize the impact of these chemicals, dismissing the studies that have shown clear, irrefutable results, studies that were carefully designed and replicated by independent scientists.

For that reason, though I don't ask you to dwell on the studies I have included here, I do want you to keep in mind that as corporate-led voices increase their attacks against those of us in the know and try to dissuade the public of these very real dangers of chemicals, it becomes even more important that you see how and where they mangle the science and misrepresent the research.

I don't intend to throw statistics, medical jargon, and scary facts at you. Instead, I want to unfold the dramatic story of endocrine-disrupting chemicals and the havoc they are wreaking on human health through an investigation into the state of our condition, as a species and as a planet, based on my access to those at the forefront of discovery—the leading minds in the fields of endocrine disruption, obstetrics, toxicology, public health, pediatrics, and more.

Since one of my areas of research entails measuring the economic costs to society from EDC exposures, you will come across some dollar amounts. This discussion of economic burden is not in any way to diminish the health dangers. Quite the opposite, actually. Because chemical companies and manufacturers often argue about the apparent high costs of safer products, it's critical to measure the economic costs of inaction so we can make a fair and proper evaluation of the tradeoffs. You will soon discover that doing nothing costs more to the public than investing in safer products and new policy regulations.

Knowledge is power, and I hope knowledge will motivate you to participate in this opportunity for momentous change, taking real steps to protect yourself and those you love.

The strongest and earliest evidence has been documented for EDC effects on the brain, and so we will start our journey there. Next we'll explore what we know about EDCs and their effects on obesity and metabolic risks. Then we'll examine the role of EDCs in increasingly common reproductive conditions.

Along the way, you will hear about the economic costs associated with diseases due to EDCs because they are large and real. Finally, you will hear the stories of people I have met in my clinical career who suffer from conditions that may be related to EDC exposures. I choose my words carefully. Nearly all of these conditions have multiple origins and factors, arising from a collision of genetic predisposition, lifestyle habits, and environmental pollutants—which is why I say "may be related" to EDCs. Some industry representatives and their scientists use the fact of these multiple factors as a way to undermine or refute the connection between chemicals and disease.

Yet, much like the probability that human activity contributes to climate change, the science has accumulated to the point that scientists are nearly 100% certain that EDCs contribute to at least one, and likely many, of the metabolic, reproductive, and nervous system disorders you're about to discover.

I trust that by giving you the knowledge about these chemicals and arming you with real, first-hand understanding of the compounds at issue and their impact, you will feel inspired to continue to grow awareness and take action. If I have done my job, you will feel more empowered, more confident in your knowledge, and more emboldened to make your voice heard. We don't need to become political operatives—far from it. Indeed, my perspective is always grounded in the best science and common sense.

My work as a pediatrician, research scientist, and policy expert is entirely focused on helping parents and others connect the dots on these massive costs so that policymakers are finally swayed. Only when we understand the astronomical burden to society, government, and the economy itself when the connection between chemicals and their effects go unchecked will our society effect true, lasting change through regulation. But it's also really important that each of us makes his or her voice heard. All of us ordinary citizens have the power and the right to influence—with our hashtags and our checkbooks. Our choices and habits matter.

This is a lot of information to take in. But understand that these chemicals are real, they are dangerous, and they are deeply entrenched in the way most Americans live their lives. And, most urgently, these chemicals are not going away without strong action from all of us.

PART ONE

THE AGE OF ENDOCRINE DISRUPTION

CHAPTER ONE

WHAT'S GOING ON?

New York City, 1962. A few dozen children between the ages of 4 and 12 are running around a concrete playground, climbing iron monkey bars, taking turns on the steel-encased swings, and riding a planked seesaw. Their shouts of glee ring in the air, with only a hint of the din of New York City traffic in the background, ships in the harbor, and exhaust from smokestacks in the city's industrial corners. It's after three o'clock and school is out. Nearly every child's mother perches on nearby benches, allowing her kids to run free and enjoy their urban oasis. Through my lens as a pediatrician, I can happily assume that most of the children have been vaccinated; that pertussis (whooping cough), tetanus, diphtheria, and polio are no longer a risk. That mumps and measles will soon be part of history. Indeed, with the 1962 Vaccination Assistance Act, the federal government began taking responsibility for protecting its youngest and most vulnerable citizens from avoidable and often fatal illnesses.[3]

I see a few kids with fresh chicken pox scars or the remains of a cold, but otherwise these kids are extraordinarily healthy compared to the generations that came before them. With widespread vaccination and the discovery of penicillin and other antibiotics, children born in the late 1950s and early 1960s benefitted from enormous strides in medical research that protected them from a slew of infectious diseases that historically had preyed on young children and contributed to high infant mortality rates. Most of the kids are of average weights and heights and demonstrate a physical and mental gusto that stems from their obvious physical health. They represent a typical ethnic and racial diversity of Manhattan, capturing generations of immigrants from all over the world.

If I were to do a closer exam of this playground melting pot, sampling and analyzing blood and urine from these kids and their moms, I would more than likely find corroboration for my observations of good health. If I were to test further and look at the markers of their emotional and social development, as well as their IQs, the kids would show consistent scores, with hardly an outlier in the group.

Now let's imagine the same New York playground in 2019: The concrete playground has been replaced by a patchwork of artificial turf and a spongy surface made from recycled tires. Gone are the metal and rubber swings, the monkey bars, and the seesaw. Instead, kids are greeted by a maze of stairs, slides, and what looks like equipment you'd find in a gym. The colors are bold—bright blue, yellow, red, green, and orange—a rainbow promising safe play. The same squeals of delight abound as the kids run and chase one another. The surrounding city is louder; gone are the whistles of the ships and the smokestacks. The streets and air appear cleaner, and when you look at the surrounding benches, the moms, dads, and other caregivers are all holding smartphones.

Looking more closely at this modern-day playground, I observe some other remarkable differences. Overall, the children are bigger and heavier. A 5-foot-1-inch boy might look 12 though he's 9. A girl with the breasts and hips of a 14-year-old is actually 8. The adults are different, too: almost half of them are overweight, a good portion bordering on obesity.

And if I were to examine these children and their caretakers more closely, doing a detailed physical and psychological screening, I would find other alarming features: At least one or two would have a diagnosis of autism spectrum disorder;[4] several kids would show significant learning disabilities; and more than 13% of the boys and more than 5% of the girls would have ADHD.[5] Many of the children would suffer from food allergies[6] and show evidence of conditions that used to affect only the elderly and infirm—diabetes, high blood cholesterol, and high blood pressure. If I looked into their futures, many of the boys would eventually have low sperm count and many of the girls would go on to develop reproductive issues, including endometriosis and infertility. In short, the basic biology and physiology of the kids have been altered in 1 or 2 generations.

What is going on here?

INVISIBLE POLLUTANTS

What has changed so radically in the intervening 57 years that the underlying makeup and health of these children and others like them have so deteriorated? What has triggered and caused the early onset of serious diseases, extremely rare just a generation ago?

Though the mix of ethnic groups might be different, the overall number of people living in New York has remained fairly constant since 1962. And, for the most part, the city has changed for the better. The city itself has survived more than one fiscal crisis and a gutting and razing of its poorest neighborhoods, but it has since flourished, rising like a phoenix, to shiny new heights. Air pollution has decreased. Access to medical care has increased. City law guarantees shelter to anyone who asks for it. Though the price of housing has climbed considerably, the city as a whole has become better at protecting its most vulnerable.

But this isn't a story about New York City. This is the story of how our physical environments in every community are currently under siege from endocrine-disrupting chemicals in our midst. This is the story of how the introduction and inundation of hundreds if not thousands of chemicals have literally changed—and by that I mean damaged—the bodies and brains of millions of people. It's the story of how many of the diseases may not yet visibly affect adults but will impact their offspring a generation or more in the future. It's a story that shows how the surface of the communities we call home—regardless of whether these communities are urban, suburban, or rural—hides a pernicious threat to the health and futures of our children and grandchildren.

Is all this simply the result of First World living? Yes and no. Many of these diseases have been attributed to sedentary lifestyles; processed, sugar-laden hyperpalatable foods; lack of exercise; and lack of access to fresh fruits and vegetables. Sequencing the human genome has made it possible to identify some of the origins of chronic diseases such as diabetes and obesity; brain disorders such as ADHD and autism; and reproductive conditions including endometriosis, low sperm count, and both male and female infertility. However, the closer we look, the more complicated the picture appears. Studies have shown that environmental exposures can modify the expression of genes (without changing the coding sequence), leading to diseases and dysfunctions. This suggests that there are other factors, so far hidden, triggering such a profound increase in these so-called lifestyle disorders.

What we now know, through rich and varied research from all over the world, is that among the hidden factors are environmental exposures to chemicals that have leached into our soil, farms, and food supply; cosmetics, hygiene products, and household furniture; and our outdoor spaces such as gardens, lawns, fields, and recreational parks. The evidence linking cause and effect is strongest for four major categories of chemicals, but we know of at least a thousand more chemicals that are endocrine disruptors. And this is an underestimate; many chemicals have not been tested and so fly under the radar of both scientists and the medical community.

The chemicals with the strongest evidence of health effects are pesticides, flame retardants, plasticizer chemicals, and bisphenols, which are used to line food and beverage cans. At first it was thought that these chemicals had to persist in the body to cause harm, like a viral or bacterial infection. Now we realize that though the chemicals themselves are often excreted within a few days, they leave lasting effects. And here is the scariest piece: the effects of this chemical contact can reverberate years later and even be passed on to the next generation. This is what I call the "hit-and-run" impact of these pernicious chemicals. They have been shown to have potent, long-lasting, life-altering effects on all of us, but especially babies and young children, whose organs are just developing, including:

- lower IQs,
- obesity,
- type 2 diabetes,
- birth defects,
- infertility,
- endometriosis,
- attention-deficit/hyperactivity disorder,
- fibroids,
- low sperm count,
- testicular cancer,
- heart disease,
- autism, and
- breast cancer.

You may wonder how such a diverse group of conditions can have something in common. They do—and it's a marker of endocrine disruption directly linked to any one or a mix of thousands of chemicals that are not yet regulated by the United States and that continue to be produced and used commercially in hundreds of products.

Although we have not yet studied all of the chemicals that exist in our homes, food supply, and environment, research supports strong if not certain links between these four groups of chemicals and diseases in at least three systems that are essential for good health: the brain and nervous system, metabolism, and reproductive functioning.

WHAT IS ENDOCRINE DISRUPTION?

Endocrine refers to our system of hormones (chemical messengers produced and used by the body). Endocrine disruption is, most simply, any disturbance in the proper functioning of hormones in the body due to a synthetic chemical exposure. Sometimes chemicals mimic the activity of a hormone by binding to receptors, causing too much of a hormone to be produced and/or released. In other cases, chemicals block the activation of a hormone or cause too little of a natural hormone to be produced or circulated in the blood. When an external chemical alters the way a hormone is supposed to function, it can cause abnormalities in cells and tissues, and organ systems such as the brain or the reproductive organs may be negatively affected, ultimately contributing to disease and dysfunction. The more we understand genes, proteins, and smaller molecules in cells, the more we realize that chemicals can change hormonal functions in subtler ways such as turning up or down the production of genes without modifying the genetic code.[7]

THE TIPPING POINT FOR ENDOCRINE DISRUPTION

Though earlier scientific publications and books had raised the alarm of endocrine disruption, a 2009 scientific statement by the Endocrine Society formally put this issue on the medical and scientific map.[8] The 17,000 members of the Endocrine Society are scientists and doctors practicing endocrinology from 120 countries, making it "the world's oldest, largest, and most active organization devoted to research on hormones and the clinical practice of endocrinology." In 2012, the World Health Organization and United Nations Environment Programme published a report documenting endocrine-disrupting chemicals as a "major and emerging global public health threat."[9,10] Three years later, the International Federation of Gynecology and Obstetrics published its own recommendations, calling for timely action to prevent harm.[11] Later in 2015, a second statement by the Endocrine Society documented even greater scientific evidence—1,331 scientific references—and concern for endocrine-disrupting chemicals and their effect on human health.[12] Most recently, in July 2018, the American Academy of Pediatrics raised the alarm about synthetic chemicals intentionally added to or inadvertently contaminating foods, urging families and policymakers to act.[13] These major international organizations are making it loud and clear: the evidence continues to grow and grow; now is the time to take concrete steps.

ACTION VS. INACTION

Federal regulators in the United States have long been aware of the connections between endocrine-related diseases and the thousands of chemicals that are still not regulated. The Centers for Disease Control and Prevention regularly surveys the US population, analyzing blood and urine from a representative sample of Americans. These surveys confirm that hormone-disrupting chemicals are quite common in the bodies of Americans. In the 2013–14 survey, 95% of adults had detectable levels of bisphenol A, and nearly all had detectable exposure to a phthalate commonly found in food packaging, di-2-ethylhexylphthalate, otherwise known as DEHP.[14]

While parents across the nation thoughtfully choose products emblazoned with the promise of being "BPA-free," they may not realize the myriad other ways they inadvertently expose their children to other harmful chemicals and hormone disruptors.

US businesses have been allowed to continue to produce, market, and distribute these chemicals under the guise of making better consumer products or benefitting food production. Industry claims that they are "safe" from under a cloud of misinformation. Consequently, we—as inhabitants of our country, continent, and planet—continue to be exposed to these harmful chemicals on a daily basis.

The European Union has begun the process of regulating many of these dangerous chemicals, but thus far, the United States has mostly been stymied. Let me share another scenario that captures this discrepancy in regulation between Europe and the United States.

Imagine you are shopping in almost any mall across the United States, preparing for a family vacation. You buy some shampoo, toothpaste, some new clothes, and snacks for your airplane ride. At the same time, a family in France or Germany is doing the same, getting ready for a much-anticipated week at the seashore. The European family goes through similar steps, buying personal care products, a few clothing items, and some food to stock their beach bungalow.

If I took a sample of the blood and urine of the two moms in each of these two scenarios, I would find a substantial difference in the number and amount of synthetic chemicals in their bodies. The average American woman would show levels of brominated flame retardants that would be considered extremely elevated in Germany or France.

Since 2003, Europe has banned 1,300 chemicals from cosmetics that are known to cause cancer, genetic mutation, reproductive harm, or birth defects.[15] In 2006, Europe enacted regulation called REACH (Registration, Evaluation, Authorization and Restriction of Chemicals) that requires chemicals commonly found in personal care products, carpeting, clothing, and food be reviewed for potential health effects and, if necessary, be replaced with safer alternatives. In comparison, the US Food and Drug Administration (FDA) has banned or restricted only 11 chemicals from cosmetics.[16]

We've also known for a while that chemicals used as flame retardants are dangerous. You may remember controversies about flame retardants being used to prevent pajamas and furniture such as mattresses and sofas from catching fire. In 1975, California law required furniture to undergo an "open-flame test" of foam filling. Materials such as polyurethane foam were tested against a small, candle-size flame for a minimum of 12 seconds. This test was easiest to pass when furniture was treated with brominated flame retardants. Only decades later, in 2013, once these chemicals were associated with damage to the developing fetal brain, the law was revised to better address fire safety without the use of chemical flame retardants.[17]

Europe never had such a requirement, and manufacturers there started phasing out brominated flame retardants in the 1990s. They were eventually banned in 2006. While there are other policy differences between the United States and Europe for flame retardant chemicals,[18] the California law resulted in different levels of exposures between the United States and Europe that have been repeatedly documented over the past decade.[19] Though these chemicals are now being phased out, there is a whole class of organophosphate flame retardants replacing the brominated ones. They won't be a focus of the book because our knowledge about these replacements is just emerging, but some relevant studies include those of Heather Patisaul and Scott Belcher at North Carolina State University, who have found accumulation of these chemicals in placentas, effects on brain development in rats, and changes in expression of genes that are crucial in lipid and carbohydrate metabolism.[20,21,22]

President Obama signed an update to the Toxic Substances Control Act in 2016 to improve the review of chemicals for safety, but the EPA is unlikely to have sufficient funding or political support to evaluate the thousands of chemicals that lack sufficient testing data.[23] In the early days of the Trump administration we've seen efforts to undermine the newly revised law. Though the United States had been ahead of Europe in limiting exposure to certain pesticides, former EPA Administrator Scott Pruitt dismissed findings from multiple carefully conducted studies as "predetermined results" and denied a petition to stop agricultural use of chlorpyrifos, an organophosphate pesticide that has been documented to damage the developing brain by affecting thyroid function during pregnancy.[24] The EPA received intense and negative outcry after making this decision, and in August 2018 a federal court ordered EPA to ban chlorpyrifos from agricultural use.

There's a very clear reason that Europe banned brominated flame retardants, chemicals used to prevent fires in furniture, electronics, and consumer appliances. The evidence linking them to effects on the developing brain is extremely strong, with studies of different populations across the world independently finding similar results and producing findings consistent with those observed in the laboratory.

You may be wondering how chemicals could conceivably create brain damage. You may question the idea that our government and policymakers could ignore or turn a blind eye to studies showing the connections between endocrine-disrupting chemicals and disease. There's no denying that there has been significant controversy as concerned parents and physicians try to wrestle with all of this information. But the fact remains that these dramatic increases in conditions that used to be rare are very real, and not one of us—or our children or grandchildren—is immune.

HOW DO WE DECIDE WHEN TO ACT?

Just over 50 years ago, in the midst of the tobacco debate, Sir Austin Bradford Hill gave a landmark lecture on assessing when and how we can be absolutely certain of causality, or a cause-and-effect relationship between any two events. He made it clear that there is no hard-and-fast rule for determining whether an association is causal because so much depends on the context of the illness and particular situation of any one individual. Hill used the example of restricting the use of a drug for morning sickness for pregnant women, suggesting that we might act on "relatively slight evidence" of harm, as "[t]he good lady and the pharmaceutical industry will doubtless survive." For an occupational carcinogen, "fair" evidence is enough for policymakers to intervene and prevent. This advice is crucial because many people (and even scientists) have misused the word *causality*.[25]

Causation is never certain; even randomized control trials used for drugs cannot always give us causality. We are constantly observing phenomena and comparing them to our human experience, adjusting our interpretations as we go. The best we can do as scientists is to lay out the information, along with all the uncertainties, and interpret the probability of a scientific phenomenon, much like climate change researchers have done. I often have to wrestle with the gray area of correlation versus causation, and like me, other scientists face the same dilemma. Often, research raises more new questions rather than providing answers on the path to understanding human phenomena. That said, the question whether to proactively prevent is not whether there is causality.

Much like members of a jury, you should formulate your own impressions. Is a preponderance of the evidence enough to act to prevent diseases that in some cases may emerge in humans only 10, 20, or more years later? Or do you need evidence beyond a reasonable doubt? My own journey as a scientist, pediatrician, and father has given me an opinion. I'll do my best to focus on "just the facts" so you can form your own.

This is the complicated work of science.

THE GOOD NEWS AND WHY WE MEASURE COSTS

Despite the direness of our situation, there is good news: We have been successful at getting chemicals out of our environment before. Take the examples of lead, asbestos, mercury, arsenic, and tobacco. It took many years and a number of pushbacks from corporations, but scientists and doctors finally convinced policymakers of the harmful effects of these chemicals. Now no one argues the fact that exposure to these chemicals produces permanent injury to the body and especially the developing brain, with lifelong consequences including reduced cognitive potential, behavioral and emotional problems, inattention, and other difficulties, often requiring support by others into adulthood. Thankfully, we now have regulations in place to protect against the well-known hazards of these chemicals.

One of the most important outcomes from studies of lead, asbestos, and mercury was understanding the enormous costs of chemical dangers to society as a whole. Indeed, while even a parent may not notice the subtle effects that environmental chemicals can have on a child's brain development, the effects on an overall population are profound. When it comes to lead, for instance, multiple studies have shown that even low-level exposures can permanently impair brain functioning and lower IQ. Now imagine for a moment if the average American lost about 5 IQ points. If the average IQ is about 100 (which means 2.5% of people fall below 70, the conventional definition for intellectual disability), a loss of 5 IQ points would mean 3.4 million more people would be intellectually disabled. That's a 57% increase, to be precise. That means the number of intellectually disabled Americans rises from 6 million to 9.4 million.[26]

From an economic perspective, studies have consistently estimated that the average (in terms of intelligence) child born in the United States will make about $1 million over his or her lifetime. Each decrease in IQ points translates to a 2%, or $20,000, loss in earning potential. Given that 4 million children are born each year, a 1-point decrease in IQ across the population of children born in a given year translates to an $80 billion reduction in lifetime earning potential and by extension overall economic productivity. This loss becomes even more substantial—staggering even—when you count not only loss of IQ but increases in ADHD, obesity, reproductive illnesses, cancers, and heart disease—all the other conditions that stem from the thousands of chemicals that still go unregulated and continue to negatively impact potential earning capacity.

When lead was phased out of gasoline in the 1970s, lead levels dropped about 12 micrograms per deciliter, which corresponded to a 2.2- to 4.7-point increase in the IQs of children born in the 2000s compared with children born in the 1970s.

To this day, because children are not exposed to lead in gasoline, annual economic productivity in the United States has been estimated to have increased by $110 to $319 billion.[27] Yes, each of us 300 million Americans gets the equivalent of as much as a $1,000 tax refund each year because we did the right thing and got lead out of gasoline in the 1970s.

At a global level, the phaseout of lead in gasoline continues to provide an economic stimulus each year of $2.45 trillion, equivalent to 4.3% of the world's gross domestic product (GDP).[28] Yes—you read that right. That's the correspondence between IQ and earning capacity. (Since lead is still used in some paints, we can still point to a 1% loss of global GDP.[29])

The scientific studies into lead contamination spurred changes in public policy to regulate the use of lead in gasoline and other products in the United States. But costs and lost economic opportunities from other chemicals, especially EDCs, still exist.

In the pages ahead, you will wrestle with this information even more—but thankfully in the context of relatable, trusted sources that will not only help you see the big picture but will also give you the tools that will encourage you to act. It's not my intention to scare you but rather to invite you to be as hopeful, optimistic, and empowered as possible.

Hearing and accepting a new reality can often be difficult and uncomfortable. Indeed, it's challenging for any of us to change our minds and open ourselves up to the possibility that we may be unwittingly endangered. I get that. Accepting the truth about harmful chemicals can trigger apprehension, doubt, and fear. But as you read the pages ahead, I want you to keep in mind that there is a lot you can do to not only protect yourself and your loved ones but also to stop the chemical rampage.

CHAPTER TWO

FOLLOWING THE HORMONAL CLUES

Allowing chemicals to infiltrate our environment—from air to soil to sea to the foods on our tables—is not new. The Sumerians used sulfur-based compounds to control insects and mites more than 2,000 years ago. The industrial revolution in Europe brought new methods of pest control to accelerate agricultural production. Worried about their crops, farmers welcomed the promises that products would protect their crops—and livelihoods—from all sorts of insects, fungi, and other natural threats to plants. While some of these products were naturally derived, they were still pernicious. For example, pesticides containing arsenic were used to keep beetles from damaging potatoes in the late 1800s. As lead came into widespread use in paints, it was combined with arsenic derivatives to make very popular lead-arsenate pesticides.

The twentieth century brought another wave of synthetic options to the pest-control portfolio. Organophosphate pesticides, for example, were first developed as human nerve gas agents during World War II. The chemicals impaired the proper functioning of the brain by blocking the breakdown of a neurotransmitter (acetylcholine) that neurons use to communicate with one another. They were subsequently recognized to be effective at killing insects via the same mechanism but at much lower exposure concentrations. The difference between humans and insects was thought to be enough to protect us back then, and the pesticide industry worked closely with the federal government and Congress to design legislation focused on "truth in labeling." The Environmental Protection Agency didn't exist in 1947 when the Federal Insecticide, Fungicide, and Rodenticide Act was passed, and so the US Department of Agriculture was the lead agency in charge. The Food and Drug Administration gained additional authority as a watchdog with the 1958 and 1960 amendments to the Federal Food, Drug, and Cosmetic Act, which banned pesticides known to be carcinogenic from foods.

Looking back, it is clear that these attempts to protect the food supply from chemicals and other compounds that can harm human health were halfhearted. The Delaney clause, which banned the use of any chemical as an additive in food if it caused cancer, regardless of the dose at which the cancer occurred, was removed in 1996.[30] In such a weak regulatory framework, much like today, corporations continued to develop and introduce thousands more chemicals into products used on more than 400 million acres of farmland.

Take, for example, dichlorodiphenyltrichloroethane, commonly known as DDT, which was first developed by a German scientist in 1873 and used to kill insects carrying diseases such as malaria and typhus. In the 1940s, another scientist, Dr. Paul Müller, discovered that DDT was an effective agricultural insecticide, and corporations began producing and selling it widely as a pesticide for crops and cattle. Products containing DDT were also used in gardens, homes, and other outdoor spaces. Soon it became clear that DDT could travel long distances and linger in the environment for years. It also accumulated in fatty tissues and was associated with the development of breast and other cancers. And yet, despite evidence that DDT was dangerous to humans, its use continued. Müller was awarded the Nobel Prize for Medicine in

1948. Although DDT was banned in 1972 by the US Environmental Protection Agency, the environment and our bodies still show evidence of exposure. Just ask Anderson Cooper, whom we found to have detectable levels of DDT as part of the CNN documentary *Planet in Peril* in 2007. More on that later in the chapter!

So if experts in scientific labs as well as corporate officers knew of the harm, why did they allow production and use to continue? I doubt they intentionally wanted to harm people, but they may have been comforted by a convenient if short-sighted argument—what I call the path of "innocent until proven guilty."

As environmentalist Will Allen points out, as early as 1929, a study published in the *American Journal of Health* showed that 29 million pounds of lead arsenate and 29 million pounds of calcium arsenate—both known to be toxic—were found in food and other materials used by people, such as cotton, coal, and building materials.[31] The postwar climate ushered in an era of "better living through chemistry," a slogan used by DuPont for nearly 50 years. The rising use of synthetic chemicals was met with enthusiasm for manufacturing appliances, making clothing and upholstery resistant to flame, and making plastics. Yes, there were a few red flags, such as deaths from apples poisoned by the arsenic in pesticides, but these events were rare, just as they are now. And synthetic chemistry offered great hope for treatments for disease and improved health. In the 1940s, diethylstilbestrol (DES) was thought to be a wonder drug for reducing pregnancy complications and losses by acting as a synthetic estrogen. More than 70 years later, we are finding new consequences in the grandchildren of mothers who took this drug.

So what's our problem? Why is safety so hard for us to ensure and practice? Why does big business always seem to have the edge over consumer protection?

VOICES ABOVE THE DIN

Thankfully, we have always been a nation that allows for, if not encourages, outspoken voices that push back on established notions, and this has been true especially in the area of environmental science. Rudolf Steiner in the late nineteenth century, Rachel Carson in the early 1960s, Theo Colborn and Pete Myers in the 1990s, and many others whom I've had the honor to know and work with have been very vocal about the dangers of chemicals. Rachel Carson broke new ground in her bestselling book *Silent Spring*, where she gathered convincing evidence of the connection between chemicals and deaths of animals; destruction of oceans, riverbeds, grasslands, and farmland; and deleterious impacts on humans. She implicated DDT in particular as an example of how "the chemical war is never won, and all life is caught in its violent crossfire."[32] But she also remained hopeful, pointing out that "[m]uch of the necessary knowledge is now available but we do not use it." And that was in 1962. Our knowledge has only grown since that time. Not surprisingly, in the wake of the popularity of *Silent Spring*, chemical corporations, the American Medical Association, and industrial agricultural companies tried to undermine Carson and the studies and assertions she made in her book. Thankfully, and as Will Allen points out in his well-documented timeline of environmental progress, in a show of support, the President's Science Advisory Committee published their support of Carson and her work.

Not long after, between 1966 and 1969, seven young women between the ages of 15 and 22 were diagnosed at the Massachusetts General Hospital with an extremely rare type of vaginal cancer. These cancers had not been seen at the hospital before 1966. While case reports of this cancer had been previously described, they occurred in much older women. This clustering of young women startled the doctors at Mass General and led to an investigation into causes. They discovered that none of the women had used any tampons or douched, only one had reported previous sexual activity, and none had taken birth control pills. This evidence compelled the concerned doctors, led by Arthur Herbst, to pull the birth records of these patients. What did they find? One feature common to all seven cases emerged: the mothers had all been given diethylstilbestrol (DES) during pregnancy. An eighth case with the same pattern was identified from another Boston hospital.[33]

To further evaluate whether diethylstilbestrol was a key difference, the researchers matched each case with four other girls based on the date and hospital of birth. Other information pulled from all of the records included birth weight, age at onset of menses, complications of pregnancy, use of other medications during pregnancy, childhood diseases of mothers and girls, history of tonsillectomy, household pets, cigarette smoking, occupation and education of the parents, cosmetic use in the mothers and girls, just to name a few. Only one factor showed a significant difference: All the girls that had the cancer were born to mothers who took diethylstilbestrol. And all the girls who did not have the cancer were born to mothers who did not take diethylstilbestrol. The chance that this difference could have occurred randomly was calculated to be 0.001 of a percentage point. (Most scientists accept 5 percentage points as a cutoff for considering an association significant and worthy of further interrogation or concern.) The findings of this now-famous DES study were published in the *New England Journal of Medicine* in 1971 and remain a milestone, setting the foundation for so much of what we know about endocrine-disrupting chemicals today. Children born of mothers who took diethylstilbestrol came to be called "DES babies."

These findings were from a population intentionally exposed

to a clinically impactful amount of a synthetic estrogen, but it prompted further concern about much lower level exposures to synthetic estrogens. DES would later be found to have effects on girls who did not in the end develop the rare cancer. We now know DES exposure is associated with breast cancer, infertility, miscarriage, and ectopic pregnancy.[34,35,36,37] Boys were not immune to the effects of DES, and so-called "DES sons" were found to have certain birth defects and noncancerous cysts in the tubes that carry sperm from the testicles.[38] Ongoing studies have also suggested an effect on obesity and cardiovascular risks decades after prenatal exposure.[39]

The studies of DES grandchildren have begun to document reproductive effects in boys born to DES-exposed women, specifically a misplacement of the urethral opening, called hypospadias, that can require corrective surgery.[40] We'll talk more about endocrine disruptors and hypospadias later, but these findings raise serious concerns about changes in gene expression (without changing the genetic code itself) that can be triggered by endocrine disruptors and passed on such that grandchildren or even great-grandchildren can suffer the adverse consequences of these exposures. We are now barely scratching the surface of the epigenome, as it's called, but laboratory studies are raising serious concerns about the future consequences of lower-level environmental exposures to EDCs that may reverberate for decades, if not longer, to come.

THE ADVENT OF THE ENDOCRINE DISRUPTION ERA

A common definition of the problems associated with hormonally active chemicals in the environment came out of the "Wingspread" meetings organized by Theo Colborn in Wisconsin in 1991. A group of world-renowned scientists had come together to recognize the broad array of chemicals that could result in deadly consequences for human health and the environment.[41] It was there and then that the term endocrine-disrupting chemical was first described. Sadly, I did not have the privilege of working with the late Theo Colborn, who earned her PhD as a grandmother at 58 and established a research organization, the Endocrine Disruption Exchange, devoted to applying the most rigorous scientific data to identify endocrine disruptors. Her work lives on as a shining example of her legacy.[42]

Around the same time, a pediatrician-scientist from Denmark, Niels Skakkebæk, was beginning to share some startling results from a study that he'd been conducting on men. Skakkebæk and his colleagues published a paper documenting an alarming decrease in sperm count in men over a 50-year period.[43] (More on this in chapter 5.) This trend has been confirmed with a larger set of 2017 data from all across the world,[44] supporting Skakkebæk's landmark findings and documenting the ongoing worsening of the effects of endocrine disruptors on reproductive function.

In 1996, Theo Colborn, John Peterson Myers, and *Boston Globe* journalist Dianne Dumanoski published the bestselling book *Our Stolen Future*.[45] Inspired by Rachel Carson's work, the book chronicled many incidents related to chemicals infiltrating our environment and hurting or killing wildlife. They directly and indirectly linked evidence of chemical destruction of animal life around the world to hundreds of different EDCs that were causing a range of effects from infertility to genetic mutations, sex changes, cancer, fatal infections, and death. This evidence included the discovery of sterile eagles in Florida in the 1940s, the disappearance of otters in late 1950s' England, a generation of infertile mink that essentially wiped out an entire population in the Great Lakes, deformed herring gulls in Lake Ontario in the 1970s, and a massive "die-off" of striped dolphins in Spain in 1992. Based on multiple analyses revealing the impact of chemicals and other pollutants on wildlife all over the world, they showed how not just one hormone system was affected but multiple. Indeed, Colborn, Myers, and Dumanoski's book brought the term *endocrine disruption* to broader attention. Myers first used the term in "response to the fact that many elements of the endocrine system were being affected, and it involved endocrine stimulation in some and an antagonistic reaction in others. It was clear that the consequences were highly disruptive for development." They also identified the parallel between climate disruption and endocrine disruption: forces messing with big systems, with many different disruptive consequences. That's what led them to suggest *endocrine disruption* as a term appropriate for the phenomenon we were grappling with. The effects on wildlife were warning signals to us as humans. While the human studies had only just begun at that time, *Our Stolen Future* ends with fingers pointing to an obvious outcome of EDC exposure: the toll on human beings.

Following the publication of *Our Stolen Future*, there were claims that the science was "exaggerated," and the authors' professional credibility was attacked. A lot of the vitriol was paid for by special-interest groups that received funding from chemical corporations that were invested in minimizing the dangers of their products. This response was not a surprise to any of us on the front lines of endocrine disruption research.

WHERE WE ARE NOW

In order to appreciate where we are now and where we need to be soon, it's helpful to see what some of us on the front lines of endocrine disruption have seen—to stand in our shoes, so to speak. I can say, without a doubt, that all of us share one thing in common: we are personally invested.

Before committing himself and his career to the eradication of EDCs, Pete Myers worked as an ornithologist. How did he make such a dramatic shift in his career? As he explains, "I'd been travelling a lot to Peru to its desert coastline to study various bird populations. Because of drastic changes in their population—90% drop in just a decade—I wanted to know what was happening to these birds," he recalled. "As we crisscrossed the region, we knew we were approaching a river because of the smell of pesticides, which were heavily used in the irrigated lands near the riverbeds. We began to speculate that pesticides might be interfering with their migratory orientation; after these birds spend the nonbreeding season in coastal Peru, they migrate to Canada and Alaska. Samples from bird brains farther north along the migration chain revealed heavy amounts of pesticides, consistent with the hypothesis." This discovery set him on a course focused on the ways in which manufactured chemicals seep into our environment and cause endocrine disruption.

Since the publication of *Our Stolen Future,* two more big issues have come to light: the connection of chemicals with obesity and diabetes—a class of chemicals now known as *obesogens*—and the transgenerational inheritance of endocrine disruption (as I described when discussing DES earlier). In other words, damages to a parent's body show up not just in their children and their grandchildren but in their great-grandchildren, too. While studies have only begun to scratch the surface of this phenomenon, we know that bisphenol A —used in can linings, thermal receipt paper, and dental sealants—and tributyltin—a chemical used to prevent microorganisms from accumulating on the hulls of ships—can induce obesity in the great-grandchildren of exposed animals, or the third generation removed from the exposure.

In a recent conversation, Pete said, "Much has changed since we wrote *Our Stolen Future*. We've learned a lot. Since then, it's clear that the types of questions we were asking were on target, and as we admitted, the science was uncertain. But it was primarily our intention to stimulate research. Now, more than twenty years later, it turns out we were right."

Unlike boisterous, extroverted Pete Myers, my friend and fellow scientist Bruce Blumberg is quiet and subdued. He was trained as a developmental biologist, the type of scientist who spends a lot of time in the lab, looking at cells of organisms in order to understand more about how they change and adapt over time and generations. As Bruce described to me, he was interested in "the sticky stuff outside of cells" and was drawn to what are known as "orphan receptors." One day he got a call from a friend, another biologist: "Hey, Bruce, have you heard about the deformed frogs in Minnesota?"

That question launched Bruce into a new direction of research that had not been on his radar: chemicals. He soon realized that the deformed frogs and other animal species that were turning up with weird mutations had all been exposed to chemicals that leached into the rivers, lakes, and oceans where they lived. He began working with a group of Japanese scientists in the early 2000s; at the time, the Japanese were investing a lot of money into endocrine disruption research. They were examining the chemicals that persist in the environment from "puddle to ocean." One specific chemical, tributyltin, was so powerful that it was reversing the sex of fish, with tiny amounts causing sex reversals in sea snails, from female to male.

Bruce began to do further tests of tributyltin, specifically on nuclear hormone receptors, his prior area of research. As a scientist, he followed the data and didn't know exactly where it would lead him. In this case, it led to a fatty acid receptor. Long story short? Bruce opened the door to understanding how some chemicals, including tributyltin, can act as obesogens in humans. "Making the human link," recalls Bruce, "may have been huge, but we never worked in humans. The link is PPAR gamma in mice and humans." This connection (a group of receptors called PPARs that influence the way calories are managed by the body) set the foundation for what is now known as the obesogen hypothesis.

Study after study published by Bruce and his colleagues have continued to show how EDC exposure triggers epigenetic alterations in gene expression, predisposing the body to produce more fat cells that lead to stubborn weight gain and obesity over time. Why is obesity harmful? Because it's linked with gallbladder disease, hypertension, coronary heart disease, and certain cancers. Children who are obese are more likely to stay obese throughout their lifetime, suffering from diminished quality of life, economic opportunity, and other negative trickle-down effects.

Bruce runs a lab at UC Irvine, where he continues to research EDCs and the link with obesity and diabetes in animals. When we met to discuss this book, I asked him what the biggest message was that he wanted to get out to the general public. Without any hesitation, he said, "Get the word out that these chemicals are real and they are dangerous to us and our children. Work with homeowners' associations. Work with schools. The moms at the local level can be a powerful force if they know what we know."

THREE STEPS FORWARD, FOUR STEPS BACK

Another crucial study that has helped us understand the effects of chemical exposures on hormones was conducted by Dr. Virginia (Ginny) Rauh, who began her career as a social worker helping poor urban children in New York. "I saw all the babies. I held them, observed them, and began to worry about them. This was in the late 1970s and early 1980s, and suddenly I was seeing many more premature babies who were not developing well. They were small."

Ginny was so concerned that she went back to school to study epidemiology to understand the effects of lifestyle on pregnancy, fetal stress, and birth outcomes. In 1998, her first funded study in children with Dr. Frederica Perera and colleagues at the Columbia Mailman School of Public Health began, and that original cohort of babies is now turning 20 years old. Her study is a good example of the roles that timing and a bit of luck have played in this field. In 2000, midway through the recruitment for her study, the Environmental Protection Agency banned residential use of the organophosphate pesticide chlorpyrifos (the same chemical Scott Pruitt decided *not* to ban for agricultural use in 2017). At the time, chlorpyrifos was widely used to combat pests in urban dwellings, and so this combination of events produced a valuable natural experiment, providing something akin to a clinical trial, which cannot be easily conducted for chemical exposures for ethical and other reasons.

Before the ban, they found decreases in birth weight and length in relationship to levels of chlorpyrifos in newborn cord blood.[46] In other words, exposure was found to be linked to smaller babies. Lower birth weights are a major concern because babies with lower birth weights are more likely to have cognitive and cardiovascular problems later in life. After the ban, as exposure levels substantially decreased, the decreases in birth weight disappeared, suggesting that preventing the pesticide exposures was beneficial. Later on, as these children aged, Ginny found that children exposed to higher levels of chlorpyrifos lost, in some cases, 3 to 5 points in IQ.[47] As mentioned earlier, while a mother may not notice a 3- to 5-point IQ loss, if 100,000 children suffer this kind of deficit, the entire economy notices. This effect is consistent with other studies showing how exposures to organophosphates are associated with decreases in cognitive function in children.[48,49] The results were significant even when researchers accounted for multiple other factors, such as socioeconomic status, and other environmental exposures, such as lead.

Ginny and her colleagues at Columbia University also followed up on the children at 7 years old with magnetic resonance imaging (MRI) studies. The highly exposed children had thinning of the frontal and parietal cortex consistent with the deficits identified in psychological testing.[50] Most recently she has found tremors appearing in the highly exposed children.[51] We'll return to the implications of these tremors and other impacts on young brains in the next chapter. But for now, keep in mind an important takeaway from Ginny's work: one of the main chemicals linked to these measurable decreases in IQ, tremors, and loss of cognitive functioning is chlorpyrifos, a chemical that the Environmental Protection Agency has deemed safe.

THE DOSE DOES *NOT* ALWAYS MAKE THE POISON

Since 1996, many scientists from diverse backgrounds and specialties have joined the ranks of those concerned about endocrine disruption. Soon after *Our Stolen Future* came out, Dr. Pat Hunt, a reproductive geneticist and, at the time, a professor and researcher at Case Western Reserve in Ohio, became an "accidental toxicologist." "It was 1998 and I was studying mouse eggs in an effort to understand why women produce abnormal eggs, a major problem in our species that causes infertility and gets worse with increasing age of the mother . . . All of the mice—both groups—had a large number of eggs with chromosome abnormalities. This happens when the chromosomes don't separate properly during the division that occurs just before the egg is released. Extra or missing chromosomes are common causes of miscarriages and genetic disorders, including birth defects. We traced the rapid increase in abnormalities to the plastic used in the mouse cages and water bottles. Apparently, the floor detergent used to clean the caging materials inadvertently damaged the plastic, causing it to leach BPA."[52]

Pat's accidental discovery made BPA a household word. "What was so frightening at the time was how rapidly the changes in eggs occurred after exposure." As a geneticist, she was interested in how low a dose she could give to get the kind of effects she was observing. This is a different approach than toxicologists take. The toxicologists who read her work said it "defied their logic" that such small amounts of exposure could cause such drastic and irreversible problems.

Two major discoveries came from studies by Pat and other prominent researchers in the field. The first was that the chemical compound BPA, which is pervasive in food and beverage cans and thermal paper receipts, triggers genetic changes specific to the reproductive system. We'll talk more about this later. The second and arguably more important discovery, which has been independently corroborated by many other researchers, has done much to undo a paradigm that had previously stood for 500 years.

Philippus Aureolus Theophrastus Bombastus von Hohenheim, or Paracelsus as he is known, was a Swiss physician, alchemist, and astrologer in the sixteenth century. Among his many aphorisms and prognostications, the one most widely quoted is "All things are poison and nothing is without poison; only the dose makes a thing not a poison."[53] Some consider him the father of the field of toxicology. Few note that Paracelsus may have made this statement to justify his use of mercury and opiates, but let's accept as a given that his judgment was clear at the time he came up with this concept.

Common sense would dictate that the amount of a toxic

exposure can reliably be used to predict its impact in a straight-line relationship. This concept is similar to the notion that we should consume "everything in moderation." For most poisons, this has proven to be true; and most if not all regulations have abided by this interpretation in deciding on safe levels of pharmaceuticals, air pollutants, chemicals used in food packaging, pesticides, and a host of other potentially hazardous exposures.

Yet, as research into endocrine disruptors has exploded, hundreds of studies have emerged showing more than a dozen chemicals with effects that disobey Paracelsus. The latest compilation of these paradigm-shifting studies by Laura Vandenberg at the University of Massachusetts and her colleagues is already 5 years old, and the evidence from the laboratory is continuing to pile up.[54] Violations of this rule are so common in the scientific literature that the director of the National Institute of Environmental Health Sciences, Dr. Linda Birnbaum, has voiced a need for regulatory agencies to reconsider their approach to account for them.[55]

The first unexpected result is that the effects may increase more rapidly at lower levels of exposure and increase less rapidly at higher levels of exposure. This is what we call *nonlinearity*. Multiple researchers have independently observed nonlinearity in the effects of lead, methylmercury, organophosphate pesticides, and brominated flame retardants on brain development.[56] Paracelsus did not have the benefit of understanding the complex molecular biology of hormones. The classic hormone-response curve is a sigmoid, or S-shaped, function because these are "high affinity receptors." Small amounts of natural hormones bind these receptors and induce important biological outcomes. The same is true for EDCs, and isn't necessarily isolated to the brain. For example, back in 2012, I, along with some colleagues, found a nonlinear association of BPA in urine levels in children with obesity.[57]

But the second anomaly—perhaps even more difficult to fathom, even for Paracelsus—is the fact that hormones can follow U-shaped, or nonmonotonic, curves in which very high exposures appear to induce less of an effect than medium doses. This is different than nonlinearity.[58] With nonmonotonicity, you can see effects rapidly accumulate at lower levels and reverse as levels get much higher. The first evidence came from the prostate, when Fred vom Saal at the University of Missouri found lower levels of DES increased prostate weight in mice, but did not see any changes in the mice with even higher levels of DES.[59] Laura Vandenberg and her colleagues found more than 600 studies of 18 EDCs with evidence for nonmonotonicity, including BPA and atrazine, an herbicide widely used in corn.[60] Again, this situation points to how difficult it is to erase misconceptions about science.

We can even explain mechanisms by which nonmonotonicity (a mouthful, I know!) can arise. Angel Nadal of the Universidad Miguel Hernández de Elche in Spain and his colleagues have examined the effects of BPA on the pancreas and observed a U-shaped function. Angel's team recently found an explanation: the shape of the response came from two different sets of receptors turned on at different levels of exposure.

The nonmonotonic effects of BPA on pancreatic beta-cells are thought to be due to two competing receptors that get activated at different levels of exposure. The "on" switch is the beta type of the estrogen receptor, while the "off" switch is the alpha type. Low levels activate the insulin "on" switch, slowing the flow of calcium into cells. The lack of calcium flow creates an electrical jolt which in turn stimulates the release of insulin granules. Higher levels activate a separate "off" switch and cancel out some of the function of the "on" switch (which stays active) by opening the same cells to calcium flow, and mitigating the electrical jolt. The lack of an electrical jolt reduces release of insulin. This combination of events gives a non-monotonic and un-Paracelsian relationship of BPA with insulin release. No exposure means minimal release, while low levels increase insulin release a lot, and higher levels produce only an intermediate change in insulin levels.

Put two straight lines on the same graph at different points, add up their effects and you have the U-shaped curve that seems unconventional.[61] The point? The dose does not make the poison the way Paracelsus envisioned it. Low-dose effects can't be predicted from high-dose experiments. Sometimes the low/medium-dose effect is stronger. Sometimes it's completely different, even opposite of what the high dose causes. Yet all regulatory testing assumes that high-dose testing will reveal what happens, if anything, at low doses.

Angel's and Pat's findings have been corroborated by other scientists and contributed substantially to our fight against EDCs and our efforts to make policy work to protect us from these known assailants. The FDA issued a federal ban on the use of BPA in baby bottles and sippy cups in 2012—a move that was considered "a small, late step in the right direction."[62,63]

Scientists face challenges on a daily basis. Pat Hunt spoke about the double-bind she was in related to her interpretation of her data and pinpointing the culprit. "We have to figure out a way to bridge the gap so that we are all on the same side." At the time, Pat was approaching the problem as a geneticist looking at the lowest levels of exposure, while the toxicologists were looking at higher levels because they were tasked with identifying safe levels of exposure. She knew in her gut that the relation between BPA and changes in the ovary was not due to some contaminant or other circumstance throwing off her results. "I've been tortured by this issue!" Pat said to me in frustration. Why? Because at first it was almost impossible for her to get clean data and convince industry representatives and even toxicologists to accept the reality that the effect of exposure to such chemicals as BPA occurs at low doses.

As of 2016, 28 states were considering or had passed legislation to limit synthetic chemicals in consumer products.[64] Those state actions spurred the chemical industry to work with environmental advocates and members of Congress to agree on updates to the major law that sets the rules for how chemicals are reviewed and evaluated by the Environmental Protection Agency. At one point, Senator Barbara Boxer accused the American Chemistry Council of effectively writing the bill.[65] True or not, there was substantial industry influence on the updates that were signed into law by President Obama in 2016.[66] While many have argued that the implementation of the new law has been undermined, it's important to emphasize that without extremely careful and hard work by a large community of scientists, the scientific data would not have been there to propel actions that protect the public.

A BIT OF CLARITY

I want to bring your attention to the tension between the scientific studies used to support the essential point of this book—the dangers of hidden chemicals on the health and welfare of all human beings, especially children—and the various groups and individuals who seek to undermine the evidence. One case in point has to do with BPA, a chemical banned from use in the production of baby bottles and sippy cups. Despite that ban, BPA is still in use, most notably in the linings of cans used to store vegetables, soda, beans, and beer, for example. Studies of BPA have engendered substantial scientific debate. As I write, the FDA released its findings from a large-scale study of BPA called CLARITY, which argues against the notion that BPA exposure is a problem, especially as a food contaminant. The FDA and industry studies paint a picture that BPA is not a problem, while the academic studies raise much more serious concerns. The FDA and industry suggest that the academic studies don't use Good Laboratory Practices, or GLP, and therefore should be dismissed. Good practices sound good, right? Ironically, GLP was developed after industry and contract laboratories were found to have falsified chemical toxicity data. Academic laboratories, which did not have problems with falsifying chemical toxicity data in the first place, rarely have the resources to comply with all the regulatory red tape associated with GLP. Instead, they rely on independent replications of their findings to ensure scientific validity and reliability.

The CLARITY study was promised to be a way that FDA, industry, and academics could get on the same page about the effects of BPA, to provide clarity in a debate that's been made out to be murky. Academics like Tom Zoeller of the University of Massachusetts came into this study with open minds and an earnest belief that the FDA would actually follow the best scientific practices, such as avoiding contamination of their unexposed animals, which have been identified in some GLP studies. Unfortunately, the flaws of GLP studies have reappeared, and the FDA decided to publicize a report on its results before vetting it through peers or collaborating with the academic scientists who worked with the FDA in good faith. Tom is a mature, steady, and calm leader in this field—you'll hear more about him in chapter 3, when we talk about EDCs and their effects on the brain. And yet he did not hold back in his criticism, pointing to industry influence, which has "kept modern science out of chemical safety determination."[67]

The FDA suggests BPA is safe by using some very old and questionable assumptions about biology. They point to tests that suggest no effects of BPA, for example, on uterine weight in mice, to argue that there are no effects on the female reproductive tract. Keep in mind that uterine weight in mice has no meaning for human health. It's also a very crude measure when exposures in early life can trigger subtle changes in the development of the reproductive tract, causing endometriosis and other forms of infertility. The FDA also dismissed any possibility of nonmonotonicity, the idea that chemicals don't necessarily have to have increasing effects as levels go up and up. Sometimes, you can see effects increase and then decrease. The FDA's own data from CLARITY suggest that there are effects at the lowest levels of exposure. They disappear at higher levels of exposure, but FDA dismisses the effects at lowest levels because they do not obey their antiquated rules about exposure and response. The reasons for effects at lower levels of exposure that disappear at higher levels are rooted in a deeper understanding of the many enzymes and receptors that can be influenced by a chemical. The reasons for nonmonotonicity are grounded in basic principles of endocrinology, something the FDA seems not to understand or perhaps does not wish to address.

I've long had my concerns about other FDA studies of BPA. One study of humans I reviewed had people collect all their urine for 24 hours to measure all of the BPA that was excreted. The FDA used the very low levels of BPA they identified to support one of its controversial perspectives about the metabolism of BPA in the human body. The average urine production of 6 liters per participant per day made my eyes nearly pop out of my sockets.[68] The average adult human urine volume is anywhere from less than 1 liter to 2 liters per 24 hours. That's a lot of drinking in the interest of FDA science, and I wondered how the participants got a good night's sleep during the study! A very straightforward and alternative interpretation of the very low levels of BPA in urine was the abundance of urine output that diluted the BPA concentration. At the very least, the outsized urine outputs throw the study's interpretation into question, which flies in the face of other studies by independent academics.

Finally, the FDA decided to publish its findings from CLARITY without waiting for the academic data, when the whole purpose of CLARITY was to figure out why there were differences between academic and industry studies in the first place. Suffice it to say that independent researchers go through stringent review before getting their studies funded by the National Institutes of Health. GLP is not a requirement for robust findings, and academic work should not be dismissed just because the work does not have the GLP label. Academic researchers submit their work for peer review, welcoming good, bad, and ugly critiques from colleagues in their field prior to publication in scientific journals. All scientifically rigorous studies should be welcomed into decisions about protecting the public, even from industry, but the FDA and the European Food Safety Authority have taken the perspective that only the GLP studies can be allowed, which leaves a lot of good science off the table.[69]

In this era of "fake news," we should be careful to interpret potential conflicts of interest in the science you might hear in the news. I would hope reporters would hold academics, industry, and government all equally accountable in this regard. Many reporters vigorously vet articles and ask tough questions that some scientific reviewers missed. I enjoy these kinds of discussions when I speak to reporters because I regularly insist on independent analyses and reanalyses of my own data before writing a manuscript for publication. Unfortunately, the media doesn't always do a terrific job of shining the spotlight on incentives to paint the scientific picture a particular way.

OUR BODY BURDEN: IT'S UP TO YOU

Early in my career, I had the privilege of working with Anderson Cooper for the CNN special series *Planet in Peril*. The TV journalist had become aware of the hidden chemicals in our midst and their connection to diseases. Testing oneself for chemical exposures is still not routine or, frankly, recommended for everyone. Most clinical laboratories do not measure minute levels of environmental chemicals with the same level of precision as research studies routinely do. As we'll see later, the data from these studies can be especially crucial in tracking the burden of disease due to these chemicals and changes in exposure that can be produced by policy change.

Anderson decided to put himself through his own experiment. He arrived at a lab in New York City to meet with me and go through what's called "body burden testing." He'd become intrigued by a family—the Holland family—who, 3 years prior, had undergone similar testing. The dad, Jeremiah, age 37, was curious to find out if any of the pernicious chemicals he had been hearing about were present in his body. They were, but what was truly astonishing was the fact that one of their children, Rowan (only 18 months), showed levels 7 times the levels that were found in his parents. Anderson wanted to know how many of the 246 chemicals able to be detected at the time would show up in his own blood and urine analyses.

We found detectable levels of over 100 chemicals, including DDT. Anderson attributed the DDT to a previous trip to Africa. To this day there remain some parts of sub-Saharan Africa where malaria is sufficiently endemic and the mosquitoes resistant to other pesticides such that DDT is still in use. (Americans who have never been to Africa also have DDT

in them because the chemical is so persistent in the environment.)

Anderson's results also suggested very high exposures to
certain phthalates known to be used in lotions, cosmetics, and
deodorants. We compared Anderson's results to the most recently available data from biological samples collected regularly from a representative sample of Americans by the Centers for Disease Control and Prevention. Anderson's level of
monobutylphthalate was above the 95th percentile in the US
population. As a representative from the American Chemistry
Council responded, the presence of such a high level does not
by itself constitute a risk to human health.[70] However, levels
below those detected in Anderson's urine have been found to
be associated with difficulty in conceiving among couples trying to have a baby. Among older adult men, these same levels
have also been associated with decreases in serum levels of
testosterone. Decreases in testosterone levels have been associated with early cardiovascular mortality, independent of
other factors.[71] Those implications are not just impacts on
people's lives but costly to the economy. We'll say more about
these findings in chapter 5.

So where are we now? Is it tenable to do body burden tests
on each and every one of us to identify and measure the levels
of chemicals in our systems? And what do we do with the information? Wait and hope for the best? Thankfully, you don't
need to spend money on these expensive tests right now. As
you will see in the coming pages, you have a lot of choices
for action, choices that I will be reminding you of throughout
this book.

Although a lot has changed in our understanding of chemicals since Arthur Herbst and his colleagues detected a cancer
cluster in Boston, we still face disbelief and collusion from
chemical industry advocates who want to quiet the truth of

the connections between chemicals and disease. In the next chapters, I will tell the stories of a dedicated group of scientists who have tirelessly and meticulously studied the effects of EDCs. We stand on the legacies of Rachel Carson and Theo Colborn and so many others. These scientists have given their minds, hearts, and bodies to seeking the truth and protecting the public. I know without a doubt that science will continue to evolve after this book is published, showing that some of what we understand now will be either incomplete or inaccurate. As scientists, that's our job: to continue to test and retest our hypotheses. What we don't have time for is to wait for more harm to occur.

WHAT'S THE PROBABILITY THAT X CAUSES Y?

Scientists and policymakers have struggled with weighing evidence as it exists now and deciding what to do in the future. The issues surrounding climate change highlight this challenge, especially since the arguments continue to play out quite publicly.

Indeed, the work we are describing here to evaluate scientific data related to EDCs builds on a framework developed by the Intergovernmental Panel on Climate Change. For each of the impacts and causes of climate change, experts came to a consensus about the probabilities using a consistent and transparent approach, based upon the strength of the evidence. These challenges are not foreign to medicine and science. Working groups have developed many frameworks to account for the state of the science using rigorous methods.[72]

For our purposes, all of the evidence collected here describes the work of experts who convened in 2014 to evaluate the evidence for EDCs and their effects on brain development, obesity and diabetes, and male and female reproductive conditions. You'll soon get to know some of those experts in the upcoming chapters, but unfortunately I don't have the space to tell you about all of their experiences and capabilities in this book.

The work of these experts appears in six articles published in the *Journal of Clinical Endocrinology and Metabolism* as well as the *International Journal of Andrology.*[73,74,75,76,77,78] The notes tell you how to find these papers in case you're looking for further details. While all of these experts are human and as humans are therefore subject to biases in evaluating the scientific evidence, this world-class group of researchers was selected for their abilities, skills, and knowledge of the relevant scientific areas. Others can debate their decisions, though all their decisions are described in great detail and with great care. And they used rigorous and careful methods throughout to get the most accurate results possible.

The six publications described disease burden and costs for 15 conditions associated with EDC exposures in Europe. My colleagues and I followed up these studies with another publication, documenting the disease burden and costs for these same conditions in the United States in the journal *Lancet Diabetes and Endocrinology*, in 2016. All of these publications were subject to rigorous vetting by anonymous peer reviewers, providing an additional measure of care to be sure we obtained the most conservative, valid, and airtight results possible.

The work of these experts was limited by time, and the science in this field is rapidly evolving. I've highlighted when I've updated the findings of these expert panels and added relevant information about other effects of EDCs that the panels simply did not have time or bandwidth to tackle back then.

PART TWO

HOW CHEMICALS HURT

CHAPTER THREE

THE ATTACK ON THE BRAIN AND NERVOUS SYSTEM

I met Michael in the newborn intensive care unit a few months after I began my pediatric residency training. He was born slightly premature, at 35 weeks, to first-time parents. He was quite vigorous at birth and probably would have gone to the well-baby nursery except that his weight was just below the cutoff our hospital used at the time. He was a slow feeder and wouldn't go home for a week.

With each follow-up visit, as Michael grew, so did my comfort in examining children and distinguishing well from unwell or normal from abnormal. Many of my colleagues had told me that this was the major challenge of internship year. Michael sailed through each visit, making up for his relatively short gestation with good, but not too much, growth, and I developed a rapport with Michael's parents.

My internship quickly rolled into residency, and Michael was growing up. At his 18-month checkup, Michael's mother, Andrea, gave me the exciting news that she was expecting the upcoming birth of a daughter and asked me if I could be the primary care doctor for both kids. As a new pediatrician, I was delighted and excited to play a role in this growing young family. During Michael's well visit, Andrea mentioned almost in passing that some of her friends—other moms in her play group—had noticed that Michael didn't seem as interactive as other kids his age.

As I began examining him, he didn't seem to be making eye contact. At first, I second-guessed myself because I had been on call the previous night, and I figured that I might be over-reacting. I asked Andrea a few more questions about Michael's behavior and temperament. He had used a couple of words by 15 months and was now using about eight words. This information was reassuring, but still I was left with residual concerns about a potential delay in Michael's brain development. These could be early warning signs of a developmental disability such as autism.

We know that early support, be it through speech or occupational therapy or other interventions, can improve long-term outcomes in children with developmental delays, whether or not the child goes on to a more definitive diagnosis of a developmental disability. I'm a good example of the positive impact such interventions can have. As an infant, I had at least two illnesses with fever that also triggered brief seizures. We know now from the Collaborative Perinatal Project that these kinds of episodes don't induce long-standing deficits.[79] In my case, I stopped talking soon after these seizures. I can still vividly remember the speech therapy sessions at the now-defunct St. Vincent's Hospital in Manhattan. Fortunately, my wife, kids, colleagues, and friends can tell you that I am not often at a loss for words now! Unfortunately for Michael, he would ultimately not completely outgrow his challenges.

THE DELICACY OF DIAGNOSING BRAIN DYSFUNCTIONS

We're fortunate now that there is a federally supported program for early intervention under the 1986 Individuals with Disabilities in Education Act (IDEA). Initial evaluations under this program are free for families that are referred to the program by a pediatrician or other health care provider, and parents can ask to have an evaluation at no charge. Early Intervention is staffed by occupational therapists, social workers, and other team members who regularly perform developmental assessments more carefully than a pediatrician can in a busy clinic setting. Under IDEA, children who are not yet in school get support and therapy that can improve their long-term outcomes.

In Michael's case, I was concerned but not yet convinced that Michael would eventually be diagnosed on the autism spectrum. As they grow, some children who have early delays in their development can catch up and go on to perform well in school and otherwise appear like a typical child.

I finished my residency just after Michael turned 2, and I left Boston to work in Washington and eventually New York City. While I don't have the details of his clinical history, I stayed in touch with Michael's family for a few years and can reassure you that with a devoted family and multidisciplinary care team, Michael's subsequent evaluations revealed a profile of autism with a normal range IQ. Some might label Michael's profile as *high-functioning autism,* though this term remains controversial for others. Generally speaking, the prognosis for the future of these children is not completely out of line with so-called neurotypical children, albeit with ongoing social difficulties and difficulties with tasks requiring motor planning, such as handwriting.

Looking back on my medical training, I realize now that Michael's situation was unusual at the time. Now, 18 years later, this story is much more common. In 2000, autism diagnoses were made in 1 of every 250 children. Today, it is estimated that 1 of every 59 children is on the autism spectrum. Among boys, the frequency is greater, 1 in 37.[80]

The available scientific data suggest increases in other neurodevelopmental disabilities as well. Eleven percent of US school-age children received a diagnosis of attention-deficit/

hyperactivity disorder (ADHD) in 2011, up from 7.8% in 2003 and 9.5% in 2007.[81] While some of these increases can be traced to changes in diagnosis, it's hard to explain away the entire increase on that basis alone.

Can I tell you Michael's autism was directly caused by EDC exposure? No, I can't. There is rarely one cause to any disease or disorder, especially those impacting the brain, and the route to causality in cases like Michael's is complicated by many factors. Was there a genetic predisposition? There was no family history to give such a hint. Not every child with autism is the same, probably due to the different parts of the brain that are affected. We know, for example, that many children with autism are diagnosed with ADHD also. And, while Michael's IQ is in the normal range, some children with autism have substantial cognitive deficits. From a research perspective, bundling children with slightly different deficits complicates finding a common signal or exposure that produces effects on specific parts of the brain. Although in my gut I believed that his autism was due at least in part to EDCs, I would be unlikely to find a chemical fingerprint from the EDC exposure. This is the hit-and-run nature of many EDCs; after they inflict their harm, they rapidly leave the body, with half of the chemical gone in a day or two in some cases, but effects that can last a lifetime.

Because of the murkiness of how brain disorders such as autism arise, scientists have come up with other, more measurable ways to connect the dots from EDCs to cognitive losses. These range from subtle deficits, not detectible in the pediatric clinic, to clinically significant disabilities that require long-term behavioral, educational, and other supports. Although male and female sex hormones play a role in brain development, the strongest evidence for the adverse effects of EDCs on the brain has been found related to the improper functioning of the thyroid hormone system, including even very subtle maternal thyroid dysfunction in pregnancy that can impair development of the fetal brain.

Thyroid hormone is perhaps best known for its impacts on growth, but thyroid hormone is also a crucial signal for brain cells (neurons, the cells that send electrical signals, and glia, which maintain the brain's architecture, among other things) to mature during fetal and early childhood development. Thyroid hormone matures key parts of the brain to take on complex thinking, decision making, the management of social behavior (the prefrontal cortex), and coordinated movement (the cerebellum).[82] Thyroid hormone also triggers brain cells to lay down the foundation so that the brain can develop and mature properly. Imagine the brain as a subway or rail system. Without the proper stimulation from thyroid hormone, the tracks are not properly connected. Once trains are allowed to run, you have the perfect recipe for a derailment, such as autism, ADHD, and other cognitive abnormalities.

Any pediatrician can tell you about the importance of thyroid hormone in newborns. In the past half-century, one of the great accomplishments in prevention is the institution of regular newborn screening, which requires pricking the heels of babies to obtain blood samples that get sent to the state public health laboratories.

Dr. Robert Guthrie initiated the newborn screening program some 50 years ago. Today, newborns are routinely screened for more than 40 disorders, including sickle cell disease, some forms of cystic fibrosis, and congenital hypothyroidism. Congenital hypothyroidism is an eminently treatable condition which is estimated to occur in 1 out of every 1,200 to 2,500 births.[83] It can result from the absence of a thyroid gland or failure of the thyroid gland to develop properly. Without early treatment, children with congenital hypothyroidism suffer severe intellectual disability.

Until about 10 or even 15 years ago, maternal levels of thyroid hormone were thought not to matter for the baby's brain development in part because the presumption was made that the thyroid hormone did not cross the placenta. But a series of studies by the late Spanish endocrinologist Gabriella Morreale de Escobar changed our thinking about the potential effects of prenatal thyroid dysfunction, proving that thyroid hormone did actually cross the placenta. A chemist by training, she was hired to measure iodine levels at the urging of her husband, a medical doctor, who was studying goiter at the time. This experience stimulated her to focus an entire academic career on the thyroid. Her work in animals found that deficits in prenatal thyroid hormone produced similar effects on the brain as those produced by congenital hypothyroidism.[84] A study of Maine children published in 1999 in the *New England Journal of Medicine* also identified cognitive deficits among children born to mothers with clinically significant low levels of thyroid hormone.[85]

The mother's thyroid hormone production is crucial because the fetal thyroid gland doesn't become fully functional until the middle of the second trimester (between 18 to 20 weeks' gestation). More recent studies have documented with remarkable consistency the broad array of consequences resulting from even subtle changes in thyroid hormone, within the clinically normal range used to evaluate pregnant women. As many as 15% of pregnant women in the United States have normal thyroid hormone levels but elevations in thyroid stimulating hormone,[86,87] which suggests a revved-up thyroid gland trying to make up for the demands of the fetus. The effects of this combination, called *subclinical hypothyroidism,* include small changes in IQ and clinically significant autism and ADHD.[88] Looking back, this combination of factors could very well have set the stage for the derailments that produced Michael's autism.

TAKING ON THE THYROID

The urgent need to prevent brain damage has spurred research on supplementing pregnant women who have subclinical hypothyroidism. In 2017, the *New England Journal of Medicine* published a randomized control trial examining the effect of giving these mothers levothyroxine (thyroid hormone) replacement. The study found no benefit for cognitive function when those children were age 5.[89] The researchers had been hopeful for some improvements in preterm birth given that some studies have associated subclinical hypothyroidism with early delivery, but there were no differences in miscarriage or preterm birth either.

There are some alternative explanations to these negative results, so it's too early to say if more precise or earlier dosing of thyroid hormone supplementation will eventually help some babies' brains develop optimally. One complicating factor in interpreting these studies is that too much thyroid hormone in pregnancy can be bad for baby, too. Some moms in the study may have been given too much, throwing off the results. Here's another analogy: Take a large group of slightly out-of-tune violins or guitars, which don't all need the same adjustment to tuning. If they get the same amount, some of them will be right on pitch, while others will be slightly off. The same goes for how we treat thyroid issues. Suffice it to say there is still debate in the obstetrics community about supplementation. At one hospital where I work, obstetricians screen moms for subclinical hypothyroidism and supplement. At another, they don't. Stay tuned, but the overarching message seems to be do what you can to keep a mom's thyroid working well in the first place, for instance by encouraging testing well before pregnancy.

Inadequate iodine remains the most common cause of

thyroid insufficiency, but there are other factors, too. Women with autoimmune conditions such as lupus can have under-active thyroid glands, and exposure to endocrine-disrupting chemicals has also been shown to disrupt thyroid function. Laboratory studies have revealed that exposure to these chemicals can produce the same patterns of cells and architecture in the brain that are produced by dietary iodine insufficiency and decreases in thyroid hormone. If you're a detective, think about fingerprints matching. Also, think about the trickiness of pointing to any one "cause."

You'll be happy to know that some thyroid-disrupting chemicals known to diminish cognitive function in children have been banned from use in the United States. Polychlorinated biphenyls, or PCBs, were used in electrical transformers and other equipment until 1977, when they were banned here. The Stockholm Convention banned their use and sale globally in 2001.[90] Now we can see levels in the US population continuing to drop; the Centers for Disease Control and Prevention biomonitoring surveys show remarkable declines. With that said, these are some of the most persistent chemicals in the soil, aquatic life, and the human body.

The "long-tail" persistence of these chemicals was shown in a landmark study. In 1997, 20 years after the ban on PCBs, Joseph and Sandra Jacobson, from Wayne State University in Michigan, carefully described differences in children born to mothers who ate contaminated fish from the Great Lakes. The most highly exposed children were 3 times as likely to have low-average IQ scores and to be at least 2 years behind in reading comprehension.[91] It was less clear how PCBs were having this effect until recently. In the past decade, studies in animals have shown that PCB exposure produces disruption in the movement of neurons to the right places in the cortex during brain development.[92] Cells also don't respond to thyroid hormone the same way, leaving parts of the brain underdeveloped and without the optimal architecture for learning and behavior.[93,94]

There is still some debate over how well the studies in humans confirm how EDCs affect thyroid functioning and go on to cause brain dysfunctions. As Sir Austin Bradford Hill said, there is no absolute threshold for causality.[95] However, you can also ask yourself this question: How much proof do you need to trust the implications that EDCs play some role in brain disorders and loss of cognitive functioning? In the case of PCBs, doctors in the 1970s were able to persuade policymakers to ban the chemicals. Why? Because they are also clearly carcinogens.

It feels like we are playing Russian roulette: Why take any chances given the evidence that chemicals contained in pesticides and flame retardants might be related to seizures, tremors, loss of IQ, ADHD, and autism, as suffered by Michael among many others? Let's take a closer look at more of the evidence.

HOW PESTICIDES ATTACK THE BRAIN

Pesticides have been around for a long, long time. Chemicals such as organophosphates were developed during World War II as chemical weapons but then began to be used to control pests, rodents, weeds, and even microorganisms. The Federal Insecticide, Fungicide, and Rodenticide Act (FIFRA) defines pesticides as "any substance or mixture of substances intended for preventing, destroying, repelling, or mitigating any pest."[96]

Why are so many pesticides still in use and still being created 50 years after Rachel Carson first shed light on their dangers? Besides the need for managing pests in agriculture and homes, a major factor rationalizing pesticide use was the theory that human brains are less vulnerable to pesticides than their rodent counterparts. At least this has been the argument for using organophosphate pesticides such as chlorpyrifos. You may recall the work of Ginny Rauh and her colleagues at Columbia University. As far back as the 1940s and until relatively recently, scientists thought that the organophosphates she measured in the children she was studying were blocking an enzyme called acetylcholinesterase, which breaks down acetylcholine into acetic acid and choline. When acetylcholinesterase is blocked by organophosphate pesticides, the organophosphates prevent pauses in nerve cell, or neuron, transmission.[97] Rodent brains are more susceptible than human brains, a reason why these pesticides were originally introduced so widely without concern. In the Paracelsian world where the dose made the poison, high levels of exposure did just that, while lower levels did not appear to, at least in humans.

A key game changer came when researchers realized that there were other effects of organophosphates at very low levels. These chemicals were shown to adversely affect the brains of animals in laboratory studies *without* inhibiting acetylcholinesterase.[98] In separate studies, scientists then discovered that the effects of certain pesticides on thyroid hormone in animals occurred at much *lower* levels than those that block acetylcholinesterase.[99] The scary part of these two findings became apparent when we realized that these lower levels of exposure corresponded roughly to the levels found most commonly in humans. And of even more concern was that researchers tied the timing of the thyroid hormone disruption in animals to a peak period of brain growth, particularly for the cortex.[100] These were animal studies, yes, but there is a long history of understanding critical periods in human brain development using animal models. The entire basis for assuming that low-level pesticide exposures might be safe in humans was revealed to have a fatal flaw. It turned this scientific field of study upside down.

It takes considerable time to confirm findings identified in animals and then see if these effects or results are also seen in humans. Why? Simply because humans have a longer lifespan than laboratory animals. Much like it took time to see whether or not Michael would actually have a developmental disability, researchers usually need to wait until children are at least 4 to 7 years of age before detecting the effects of chemical exposure in pregnancy on IQ. And this assumes you have funding for a human study the instant the suggestive laboratory studies are published, which is overly idealistic. Add in some time up front to apply for funding. On the back end, throw in some additional time for statistical analyses, debating the results, writing them up, and navigating the tough peer review process. You can now see why there often is a substantial lag between laboratory studies and their human counterparts.

The EDC Disease Burden Working Group I mentioned at the beginning of the book had different subgroups, with one specifically focused on effects on brain development. The neurodevelopmental experts I brought together in 2014 looked at three carefully conducted long-term human studies, one of them by Ginny Rauh and her team at Columbia, another in New York City, and one from a farmworker community in California.[101] They all gave the same interpretation: As exposure to these insecticides during pregnancy went up, the child's IQ went down; the only question was by how much. The data suggested that for each tenfold increase in pesticide level, the effect ranged from 1.4 to 5.6 points.[102]

Ginny's study was particularly important because of the policy change that occurred during recruitment, as I mentioned in chapter 2. The pesticide chlorpyrifos was banned for household use, resulting in lower levels of exposure. Children born before the ban had lower birth weight and length; the effect was not seen in those born post-ban. Measures of birth weight and length are strong predictors of later brain development. In the *Proceedings of the National Academy of Sciences,* Ginny and her team published the stunning differences in the brains of the highly exposed children at age 7.[103] The thinning of the frontal and parietal cortex was consistent with the deficits identified in psychological testing. Would we have seen these kinds of changes in Michael's MRI? We'll never know for certain, but it's distinctly possible.

Obviously, for ethical reasons, we don't run randomized control trials of pesticide exposures. For chemical hazards, we have to rely on studies in which we observe humans and examine health effects of the pesticide levels they happen to be exposed to in their everyday lives. A further difficulty is that the relationships of exposure and effect don't follow straight lines. And the associations are just that, falling short of proving causation by themselves. But when the reality of these so-called observational studies is consistent with the results of animal experiments (where you can control many other factors), the two add up and can be remarkably helpful and valuable for interpreting potential cause-and-effect relationships. The experts I assembled in 2014 pulled together criteria from multiple reputable and established sources to weigh the evidence and estimate the probability of causality for prenatal organophosphate exposure and its adverse effects on cognitive function. Not coincidentally, some of these methods were developed by the Intergovernmental Panel on Climate Change to deal with similar challenges in weighing the evidence for climate change.

When the team presented their results, I was absolutely shocked. They estimated at least a 70% probability, and some members supported a higher percentage, nearly 100%. The message was clear: the evidence for organophosphate pesticides' effect on the brain by disrupting thyroid hormone is nearly as convincing as the evidence for lead poisoning.

We started calculating the implications for the population

of kids born in Europe in 2010. If a child comes back from school with an IQ point loss, a parent might not notice, but such losses at a population scale stand out. We extrapolated the consequences to Europe, using the most representative exposure data we could find. The most likely scenario suggested that no child in Europe suffered more than a 5.3-point IQ loss. Due to the slightly different results in the human studies, there are uncertainties, and so we made a range of estimates. In the best case scenario, the most-exposed kids lost 1.7 IQ points, and in the worst case scenario, the IQ loss would be 7.0 points.[104]

Mothers notice nearly everything, but they might not notice that subtle a difference in a child's intelligence. Yet the economy definitely notices. As I described in chapter 1, each IQ point lost translates to a 2% drop in lifetime earning potential.[105] This loss of potential is due in part to less participation in the workforce but could also be the result of lower wages. If you keep in mind that the average person in the United States makes $1 million over a lifetime, then an IQ point on average is worth about $20,000. Adjust for exchange rates and differences in purchasing power parity, and the cost of an IQ point in Europe ranges between €5,000 and €25,000. With 4.5 million Europeans losing anywhere between a third of an IQ point and just over 5 each, we estimated the total lost economic productivity to be €125 billion. Then, add in the fact that these losses shift some 59,300 children into the intellectual disability range, with additional educational and other costs, and you add another €21 billion to the cost.

The final results boil down to a 70% or more probability that a toxic environmental exposure costs €146 billion per year (which is equivalent to $194 billion per year using the current exchange rate). The list price for a brand-new Boeing 787 is $194 million. Imagine a 70% chance that one of those planes is stolen *each day*. That would not cost as much as organophosphate pesticide exposures do in Europe. It's a 70% chance of 1,000 Boeing 787s being stolen in a given year. Assuming exposure stays the same in the absence of change in policy or behavior, we repeat the cycle of stealing planes each year, as a new cohort of children is born with the same consequences.

So what does this all mean?

The good news is that the EPA ended the use of the organophosphate chlorpyrifos in households in 2000.[106] This was a direct consequence of the 1996 Food Quality Protection Act, or FQPA, which added a protective safety factor for children that reduced the allowable levels of organophosphate pesticides. These events substantially reduced organophosphate exposures in US children.[107] In an example of how environmental health issues don't need to be partisan, the FQPA unanimously passed both houses of Congress.

In 2016, we published an analysis in the Lancet focused on diabetes and endocrinology and found a much smaller cost of organophosphate pesticides in the United States—$45 billion—compared to Europe ($194 billion). A Boeing 787 being stolen every week is better than every day but still might get President Trump's attention. Policies change exposures, exposures contribute to disease, and diseases are costly to our economy, too. In this case, America is literally smarter—and richer—than Europe because of environmental regulation that protects people.

I did warn you about some bad news. Unbelievably, the US Environmental Protection Agency (EPA), then led by former administrator Scott Pruitt, in April 2017 denied a petition to revoke all food residue tolerances for chlorpyrifos, calling it "crucial to US agriculture" and to "ensur[ing] an abundant and affordable food supply for this nation and for the world."[108] The petition in front of the EPA was a proposal to ban all uses of chlorpyrifos because of the proven risks to women and children, as well as farmworkers. What did Pruitt say in his press release? He suggested that the harmful findings were "predetermined" and therefore any request to shape policy was biased. His department also fell back on the argument that chlorpyrifos was the only effective choice for some crops. This is patently untrue, and a federal court later agreed that EPA had acted illegally by ignoring agency scientists who had warned that chlorpyrifos was harmful.

This "feed the world" argument is growing thin. A recent meta-analysis comparing the yields of conventional to organic crops showed that the two agricultural methods may be equivalent under good management practices, particular crop types, and growing conditions.[109,110] Let us for a moment assume that there are no alternatives to chlorpyrifos and that it is needed to sustain the global food supply. At the very least, there are serious tradeoffs to consider in such a decision, with a substantial number of children impaired from being able to fully contribute to the future of the global economy. Administrator Pruitt's statement on chlorpyrifos is mum on this point.[111]

WHAT YOU CAN DO NOW

How can you reduce your exposure to pesticides? For starters, you can purchase and eat organic foods—fruits, vegetables, pasta, rice, milk, cheese, meats, and poultry. We are lucky to live in a world where organic foods are now increasingly available. Leafy greens and vegetables are the foods for which eating organic matters the most simply because you are eating the part of the plant on which the pesticide is sprayed. No washing method is 100% effective for removing all pesticide residues.

But before you go rushing out to your grocery store, let

me share a few details about how eating organically can make a difference. In one study from 2006, Chensheng Lu and colleagues, who were then at Emory University, followed a group of elementary-age children for 15 days, giving them only organic foods. The result? The scientists showed that "an organic diet provides a dramatic and immediate protective effect against exposures to organophosphorus pesticides that are commonly used in agricultural production." They also could confidently conclude that the higher prestudy levels of pesticides in the children's urine were due to exposure through the foods they were eating.[112] A more recent study, in 2015, by Asa Bradman and colleagues at the University of California at Berkeley, also confirmed the effectiveness of an organic diet intervention in low-income urban and agricultural communities. The success of the study speaks to the value of targeted efforts to buy organic for the fruits and vegetables of greatest concern. Indeed, traces of organophosphate metabolite as well as the herbicide 2,4-D (which has been associated with non-Hodgkin's lymphoma and sarcoma, a soft-tissue cancer) both dropped.[113]

Eating organically makes an immediate, measurable difference. Just to be clear, I'm not talking about nutritional differences—these have been studied without convincing results. When you purchase a food that is labelled "organic," you can be confident that it doesn't contain any genetically modified organisms (GMOs). And though there still is some scientific debate related to the safety of GMO foods, I caution against their use, especially since some genetically modified crops have been treated with such herbicides as glyphosate. To be clear, genetic engineering in and of itself need not be an ill. It's been disturbing to see the GMO debate become a genomic versus nongenomic debate. The tradeoffs are much more complex than how they have been framed by many who have financial gain at stake.

FLAME RETARDANTS

Another group of chemicals with overwhelmingly convincing evidence for their effects on prenatal thyroid hormone, and ultimately brain development and functioning, are loosely generalized as flame retardants. Flame retardants are present in furniture with foam inserts, synthetic fabrics, carpeting, and flooring. Within this group, the chemicals of greatest concern are comprised chiefly of carbons and bromines (called organohalogens) and particularly the ones called polybrominated diphenyl ethers, or PBDEs for short.

The point of this book is not to turn you into a chemist. With that said, it's easy enough to use your favorite Internet search engine to search for diagrams associated with PBDE and thyroid hormone. The chemical structures are remarkably similar. One difference is there are bromines in PBDE and iodines in thyroid hormone, but these two elements are in the same column on the periodic table, with some similarity in their atoms. Function in the human body often follows chemical structure. Look at the way pharmaceuticals are created. Drugs are designed by matching chemical structures that will fit in receptors, causing a cascade of effects in the human body. Synthetic chemicals are not made that way—they are designed for the properties they provide to materials, in this case making them flame resistant. But even though these synthetic chemicals are not designed with the human body in mind, they can inadvertently create conflicts when they have structures that are similar to naturally occurring hormones that induce biological effects. And this is the case for many flame-retardant chemicals.

The neurodevelopmental experts from the EDC Disease Burden Working Group I mentioned earlier put together a table of 20 laboratory and animal studies of flame-retardant chemicals, which showed a consistent (if not perfect) pattern.[114]

PBDEs interfere with the binding of thyroid hormone to its receptor. PBDEs can also disrupt thyroid hormone metabolism, limiting the impact of available thyroid hormone. Additionally, some of the research suggested effects on the animal brain independent of thyroid function. This means that these chemicals likely interact with other systems and pathways that are yet to be identified.

We reviewed the results of four other studies examining levels of PBDEs in mothers' blood or babies' umbilical cord blood, and the children's intellectual capacity as measured with IQ tests.[115,116,117,118] Three US studies showed negative effects on cognitive function as PBDE levels increased; we'll discuss what the studies have found for autism and ADHD later. The researchers carefully controlled for other factors such as socioeconomic status and other environmental factors that could explain away the observations, but the associations remained.

One Spanish study showed less convincing findings, revealing the implications of an important difference between policies in the United States and Europe.[119] Given the California requirement to use flame retardants that did not exist in Europe, levels in the Spanish study were much lower than in the US studies. It's quite remarkable that the Spanish study detected a trend toward lower cognitive and motor function, because the lower levels likely made it difficult to detect effects on the brain in the first place.

The neurodevelopmental experts on the EDC Disease Burden Working Group decided that there was very strong evidence (70% or more likely, similar to what was found for organophosphate pesticides) that linked PBDEs as the probable cause of the observed decreases in cognitive function in children.

PBDEs are a good example of the implications of policy differences for increased disease and disability and ultimately costs to society. More kids in the United States are affected by prenatal PBDE exposures than in Europe, and the IQ losses are greater. The costs of PBDE exposure are huge—$266 billion in the United States compared to $13 billion in Europe.[120] The sad reality to add to this picture is that there are racial and ethnic differences in exposure to these chemicals. We have just completed a study documenting that non-Hispanic whites bear only 54% of the burden of PBDE exposure, even though they comprise 66% of the population born in 2010. African Americans and Latinos bear a larger share of these consequences and costs.[121]

GET TO KNOW YOUR CHEMICALS: PBDES

This class of chemicals, which are derived in part from bromine, a poison, has increased in use since California enacted legislation in the 1970s to protect against home fires. PBDEs can be found in plastics used for furniture (sofas, chairs, and mattresses), electronics, wire insulation, foam used in car seats, and carpets. PBDEs and other flame-retardant chemicals are also added to children's toys, clothing, and other children's products.

WATCH OUT FOR HIDDEN CULPRITS

You're probably getting the picture that there are numerous ways that the thyroid hormone can be affected by chemicals and trigger one or more brain dysfunctions. Here's a rundown on several other chemical culprits that can have additional, downstream effects on the brain.

• Hidden chemicals in products used to package food—

such as plastic wrap, baggies, and to-go containers—are a potential source of thyroid disruptors. In particular, **perchlorate,** a chemical used in the production of rocket fuel, missiles, fireworks, flares, and explosives, is also used to prevent static cling on plastic and paper packages. It interferes with the uptake of iodine needed for thyroid hormone production, as does **thiocyanate**, a contaminant found in cigarette smoke, and **nitrate**, which is used in fertilizers. A study examining pregnant women who were already hypothyroid found high perchlorate levels in mothers to be associated with lower IQs in their children.

- You have already heard a lot about **bisphenol A (BPA).** This synthetic chemical is used in the lining of food and beverage cans, as well as thermal paper receipts, even though it's been banned from baby bottles and sippy cups. BPA is best known as a synthetic estrogen; in fact, it was considered as a pharmaceutical agent for pregnant women to prevent miscarriage but was not as potent as DES (you know the DES story from chapter 1). It can disrupt thyroid function and inhibit thyroid hormone binding during development of the cortex, the part of the brain involved in so many functions unique to humans. You may have seen a lot of plastic water bottles and other containers with the "BPA-free" label. As BPA-free has come into vogue, so have a bunch of replacements: BPP, BPF, BPS, BPZ, and BPAP, just to name a few, which pose similar if not worse threats. It's best to avoid hard, polycarbonate plastic containers altogether.

- **Phthalates** are another source of thyroid inhibitors and disruptors. This group of chemicals is used to make plastics soft, such as in food packaging, and to enhance scents and smells of personal care products such as lotions and cosmetics. These are a very diverse group, which we'll talk about in the next chapter as possible contributors to obesity, metabolic, and reproductive effects. Some of the phthalates used in lotions and cosmetics act to block the effects of the male sex hormone testosterone, while other phthalates used in food packaging and flooring can act as estrogens. In laboratory studies, we can focus on a single phthalate or other chemical exposure, and indeed phthalates found in food packaging can influence expression of genes that propel production of thyroid stimulating hormone. These chemicals can "hit and run," leaving no specific fingerprint of the exposure to allow us to deduce the kinds of effects they have on the brains of developing young children.

- Perfluoroalkyl substances, or **PFASs**, are molecules known for having multiple fluorine atoms in their structure, which give nonstick characteristics that are attractive for their use in textiles, furniture, and cookware. Perfluorooctanoic acid (PFOA) is one of the "long-chain" PFASs and is used in Teflon. These chemicals interfere with thyroid hormone binding. Though some human studies have shown potentially harmful effects on brain development, the evidence is stronger for their effects on fetal growth and birth weight. Other harmful effects in humans have been documented in communities near manufacturing plants that produce long-chain PFASs. Eventually, chemical companies negotiated a phaseout of these PFASs, and PFOA levels in people and the environment have dropped. But you still see water contamination from previous use of PFASs all across the country. For example, in Hoosick Falls, a community in upstate New York, two major manufacturers have been sued after the drinking water supply was found to be contaminated with PFOA.

- One of the replacements for PFOA is a chemical called **GenX.** No, that's not referring to Generation X'ers (I'm one of those and am not disaffected or directionless as some have described us). GenX is the product name for a chemical that was developed to replace PFOA. GenX is used for the production of common household products, including Teflon used for nonstick pans, firefighting foam, and common outdoor fabrics. Chemical plants producing GenX in North Carolina and Ohio have become the focus of a lot of public attention of late because of nearby water contamination. Early laboratory studies from the Netherlands and Sweden have found GenX to produce similar or even greater effects on fetal growth and birth weight as PFOA.[122,123] The manufacturers claim that these and other short-chain replacements for long-chain PFASs get removed from the human body a lot faster and therefore are less harmful. Dr. Linda Birnbaum, Director of the National Institute for Environmental Health Sciences, could not disagree more, noting that "every PFAS that has been studied is causing problems." She continued, "Even if they have a shorter half-life, if it has a half-life of 30 days, it's going to build up in your body."[124]

- **Organophosphorus flame retardants,** which are being used to replace PBDEs, may present a similar problem. We call this problem "regrettable substitution" or "chemical whack-a-mole." It's a byproduct of the regulatory framework that assumes chemicals are innocent until proven guilty. Once a chemical is identified to be a problem, minor changes in a chemical structure are implemented that don't change the manufacturing much, but they reset the way it is treated under the law. Because these chemicals are added during the manufacturing process and are not bound to the products in which they are used, they are widely found in dust in homes, offices, and even cars. Not to make you self-conscious, but on average, we touch our faces two to five times each minute, unconsciously ingesting the chemical dusts we come into contact with. Laboratory studies have already suggested effects on the developing brain.[125] These chemicals are relatively new on the horizon, so more human studies will be needed to confirm whether these new flame retardants are as problematic as the ones they replaced.

CLINICAL CONSEQUENCES OF EDCS: ADHD

Effects on IQ are not the only ways chemical exposures can affect the growing brain. As a pediatrician, I could not be more emphatic about the need to prevent exposures that contribute to conditions like ADHD and autism. Studies of the relationship between EDCs and ADHD are challenging in part because ADHD has two components—inattention and hyperactivity. No child—or adult—with ADHD has the same amount of one component or the other. Also, the diagnosis of ADHD requires observation of inattention or hyperactivity in more than one setting. The diagnosis is typically made by a pediatrician, child psychiatrist, or other care provider, which leaves open the possibility that the human factors that go into diagnosis can get in the way of looking at diagnosis by itself. The alternative is to measure inattentiveness and hyperactivity with scales that researchers have honed, but this relies on a parent's interpretation.

As a result, it's very hard to decipher all the factors that can cause ADHD. However, there has been consensus on a number of very solid links. In particular, flame retardants such as PBDEs and organophosphate pesticides have both been investigated. Two large-scale studies found associations of ADHD with PBDE exposures, though one found stronger associations with the degree of inattentiveness and the other with the degree of hyperactivity.[126,127] A study of organophosphates found associations with the diagnosis of ADHD, but not the number of ADHD symptoms.[128] If organophosphates were causing ADHD, you would expect chemical exposures to be associated with both the presence and severity of ADHD. We do know some things from studies in mice: When researchers remove one of the thyroid receptor genes, the mice exhibit ADHD-like behaviors[129]; the same behaviors show up with lower levels of thyroid hormone.[130] We also know that EDCs can induce ADHD-like activity in rodents.[131,132] All that said, the EDC Disease Burden Working Group advised — and I agree — that the probability of causation is moderate, more like a coin flip.

Why don't we have more conclusive answers?

Under the current regulatory environment, chemicals are produced and used with very little data about their safety. Instead, safety is assumed. The probability of adverse effects is presumed to be zero. Think back to Paracelsus. We assumed the dose always made the poison, and here we are 500 years later realizing that we had it all wrong. Do you want to take a chance and risk exposure to chemicals that more than likely set the stage for ADHD? Even if the evidence for effects on attention are more moderate, the data related to the effects of PBDEs and organophosphates on cognition are strong and suggest a need for action. We estimate that 4,400 children each year might develop ADHD because of EDCs. The costs of EDCs contributed by ADHD are a shade under a billion dollars per year, much smaller than the costs of EDCs on IQ despite the fact that the costs per affected child are much larger. That number may seem small, but keep in mind that not every child suffers from ADHD, and fewer still suffer ADHD due to EDC exposure. Even so, the consequences of ADHD are much more substantial compared to the effect on each child who loses one or more IQ points. Also keep in mind that these chemicals may be likely making all children more inattentive and/or hyperactive, with diagnosis occurring in an increasingly larger number of children whose symptoms trigger a diagnosis.

AUTISM

We began this chapter talking about my patient Michael and autism. Although autism diagnoses have increased, it is still not as common as ADHD or asthma. From a research perspective, this is a problem. If you are trying to examine the influence of any risk factor—nutritional, genetic, socioeconomic, or otherwise—on the incidence of autism, the statisticians will tell you that you need to study perhaps 10,000 children or more before you can feel confident interpreting results with a high likelihood of picking up the important associations. In smaller populations, you might find no statistically significant association, but you are unlikely to say there was no association at all. Autism has long been known to affect boys more than girls, and EDCs can affect boys' sex hormones differently than those same hormones in girls. That makes the clues in this detective story even harder to find.

A New York City–based study found that phthalate exposure in pregnancy contributed to increases in a scale used to evaluate autism,[133] but most of the kids in this study did not have autism. Another study based in Cincinnati found four other EDCs to be associated with changes in these scales.[134] These studies are both small but suggestive. Measuring the impact is harder. There are ways to convert increases in scales to estimate potential increases in autism, but interpreting these takes great care. We estimated that nearly 10% of autism could be due to EDCs, with a possibility of a lower percentage, closer to 2%, and we couldn't specify which EDCs due to the lack of data.

The good news is that a very large and exciting national program called the Environmental Influences on Child Health Outcomes (ECHO) is pulling together data from 50,000 children across the United States to understand the effects of EDCs and other environmental exposures. For autism, ECHO will make a lot of progress in working the important questions that are left. (For more information, check out www.nih.gov/echo.)

So, with all this uncertainty, what is the best course of

action? There's a lot we can do now while we're figuring this out. Surprisingly, it's not all about changing exposures—eating a diet with sufficient iodine can go a long way, as studies have indicated. Avoiding canned foods and not microwaving plastic are on that list, too. Studies have documented that eating out is associated with higher levels of phthalates as measured in urine. We've talked a lot about probabilities, which may scare some. Another way to think about these is to ask: How willing are we to gamble with our health?

WHAT YOU CAN DO NOW TO LIMIT EDC EXPOSURES THAT AFFECT THE BRAIN

- Choose naturally flame-resistant materials like wool. Upholstered items with the TB117-2013 label do not require flame retardants. Clothing that fits snugly meets flame-retardant standards without added chemicals because they are tight enough such that no stray sleeve can catch fire. It also increases flame resistance because there's no extra air between the fabric and the skin to promote the spread of a fire.
- Use a wet mop to get rid of chemical residues, and regularly open windows to recirculate the air in your home.
- PBDEs can accumulate in animal fat. Eating a plant-based diet avoids these exposures.
- Avoid canned foods and foods packaged in hard polycarbonate plastics (typically with the number 7 in the recycling triangle on the bottom).
- Say no to that coated receipt at the cash register.

CHAPTER FOUR

METABOLIC MIX-UPS: OBESITY AND DIABETES

You may recall from the playground scene in chapter 1 that a good portion (almost 35%) of American children are now considered obese or significantly overweight. If you compared riders on the New York City subway in 1962 to those today, you'd see big differences among the adults, too. In 2016, almost 40% of American adults were obese. Obesity is defined as a body mass index (BMI, the measure health care providers use to evaluate the degree of fat relative to lean body mass, based on your height and weight) of 30 or higher. In 2016, another 30% were overweight, which is defined in adults as a BMI between 25 and 30.[135] Americans with a BMI under 25, considered normal weight, are now in the minority.

We are in a new (ab)normal. Obesity or overweight is not a disease by itself, but both are associated with heart disease, diabetes, liver dysfunction, stroke, and certain cancers. Obesity is also costly, resulting in $190 billion per year in weight-related medical bills.[136]

Nothing I will tell you in this chapter should change your thinking about the leading drivers of the epidemic—unhealthy diet and physical inactivity. What you may not know, however, is that changes in diet and physical activity cannot fully explain the increases in obesity. In theory, if you eat 50 more calories per day than you expend in exercise, you would gain an additional 5 pounds as fat per year. You can imagine then that the changes in obesity in the United States can be explained completely by increased calorie intake and less physical activity, right? If only it were that simple. The problem is that obesity is not simply a matter of calories in versus calories out. The same surveys from the Centers for Disease Control and Prevention where we get all that data about chemical exposures also measure calorie intake and physical activity from nationally representative samples of Americans. Data from those surveys tell us that between 1988 and 2006, frequency of physical activity during leisure increased, not decreased, 47% in men and 120% in women. In fact, when researchers looked at two adults with the same calorie intake and level of physical activity, one from 1988 and the other from 2006, the adult in 2006 had a BMI 2.3 points higher![137] That would be enough to shift someone halfway from overweight to obese.

Diet is not simply a matter of calorie intake. Dietary composition matters, too, and the composition of the American diet has shifted toward sugar. Dr. Robert Lustig of the University of California at San Francisco has called high-fructose corn syrup a poison.[138] I'm not in disagreement, nor is anything I am about to tell you inconsistent with this perspective. Lack of sufficient sleep and poor sleep quality may also play a role in exacerbating weight gain.[139] Access to healthy, clean foods and daytime physical activity is also important to maintaining a healthy metabolism.[140]

The obesity problem is not simply a reflection of more technology time, sedentary lifestyles, and a diet made up of sugar-laden processed foods, however. Carefully designed, peer-reviewed studies have drawn an increasingly convincing link between obesity and type 2 diabetes and prenatal and early childhood exposure to pesticides, bisphenols (such as BPA), and plasticizers such as phthalates. Other studies have suggested that adults may gain weight or develop diabetes in response to their exposure later in life.

In case you didn't read the box in the last chapter, PFOA is a chemical that until recently was used in nonstick pans and clothing (to make it stain resistant). In a very recent study published in the *Public Library of Science Medicine* journal, levels of PFOA in people's serum were associated with weight regain among people who had successfully lost weight through better diet and physical activity. Higher PFOA levels were associated with slowing of resting metabolic rates, which can cause weight gain.[141]

You may have heard some people referring to themselves as having a "slower metabolism" than others. Our tissues have different energy-use rates, just like cars have different fuel-usage rates. Add up the energy usage rates of all the parts of the body before considering exercise or other forms of physical activity, and you have a resting metabolic rate. It may be that chemical exposures influence the fuel efficiency of our bodies in such a way that the same calorie intake can induce more (or less) weight gain.

You are already familiar with organophosphate pesticides and their impact on the brain and nervous system. Now you will see how some other pesticides and other common chemicals that we all come into contact with on a daily basis mess with the way the body is meant to metabolize food and maintain homeostasis. Children are the most vulnerable, but adults are also susceptible to chemical obesogens, metabolic disruptors, and cardiovascular risks.

THE THRIFTY PHENOTYPE

Michelle is a bright young girl I first met in the primary care clinic when she was entering first grade. Her parents had arrived from Mexico a few years before she was born and put down roots in El Barrio, a neighborhood in northern Manhattan just east of Central Park. Her entire family was extremely proud of the local charter school she was attending, frequently bringing her to appointments still in her full school uniform.

On reviewing growth charts at her annual well-child visit, I noticed she had a prototypical profile of a child at risk for obesity. She had been born on time, but her birth weight was low for her gestational age, just below the threshold for being small for gestational age, or SGA. SGA poses immediate risks for the newborn, such as low blood sugar levels, but these are readily manageable with careful monitoring and clinical intervention. In infancy, her weight rapidly caught up, and by her 2-year-old checkup, Michelle's weight-for-length hovered around the 95th percentile for a girl of her age. After age 2, her BMI danced between the 85th and 95th percentiles, the first being the cut point to be labeled overweight, equivalent to a BMI of 25 in adults, and the other the threshold for being called obese, equivalent of a BMI of 30 in grown-ups.

Michelle is a good example of what we call the "thrifty phenotype" problem, which was first described by the late Sir David Barker.[142] During World War II, between November 1944 and May 1945, the Dutch experienced a severe famine in which daily rations were as low as 400 to 800 calories. Babies born to women who were pregnant at the time experienced severe growth restriction. Though they appeared otherwise fine at birth, decades later they were found to suffer obesity, diabetes, high blood pressure, and early coronary artery disease, dying earlier than others who were unaffected by the famine. Barker developed the "thrifty phenotype" hypothesis to explain these results, arguing that the fetus adapts to nutritional and other environmental insults to survive, optimizing the use of every calorie it receives. Once out of the womb, in a freer nutritional environment like that experienced after World War II, the child maintains this "adaptive" response, storing excess calories as fat or in other ways that present problems even many decades later.[143] The fetal stress is thought to modify the genome, not by changing it directly but by putting signals on specific genes that code for proteins that play key roles in lipid and sugar metabolism.[144] Over the 50 years since the Dutch Hunger Winter, we have learned that even subtler environmental stresses can restrict fetal growth and produce the same kind of long-term effects on obesity and cardiovascular health.[145]

Fortunately, the pediatricians who had preceded me in caring for Michelle realized her risk profile and had already proactively coached her parents to avoid junk foods and make sure she exercised daily—actions to offset this metabolic propensity to hold on to weight. Michelle was lucky in that her charter school had a terrific soccer program, which she enjoyed. She also played and practiced soccer with her father on weekends. With that said, balancing calories in and calories out can be challenging in some neighborhoods in New York City, even with Central Park as a playground. The notion of "food deserts" speaks to the reality that healthy eating options may not be available close by or may not be financially within reach. Michelle's parents wanted her to have a healthy mind and body, but even with Dad working both full-time and part-time jobs to keep a healthy income flowing, gentrification in East Harlem has been pushing rents higher, making it harder to go to the supermarket rather than the bodegas, which are known to sell fewer fresh fruits and vegetables.

But I didn't think diet was the whole problem for Michelle. What was different about Michelle that made controlling her weight so difficult when others were able to keep a healthy weight? Was Michelle fighting an uphill battle because of her growth restriction in the womb? If it wasn't her diet or physical activity that was driving metabolic disarray, what in her environment was beyond her parents' control?

Around this time, in 2009, I was beginning to become aware of chemical obesogens. The studies mostly examined effects of bisphenol A and phthalates on laboratory animals. I was working hard in another large study of US children called the National Children's Study (NCS), which preceded the one I am working on now. A paper I published in *Environmental Health Perspectives* that year, with fellow researchers from the National Children's Study, talked about the progress Arkansas had made in reducing obesity among school-age kids, and the role of diet, physical activity, and other environmental factors in childhood obesity. In that article, we also wrote about how DES was found to trigger obesity in animal studies. There were other suspected chemicals, but they had not been examined in humans, and the NCS was planning to fill a big scientific void by measuring exposures and then tracking growth that followed to see if there was a relationship.[146]

Unfortunately, the NCS was shut down in 2014 by the National Institutes of Health because its approach to recruiting families was found to be too costly and administratively complex.[147] In the interim, smaller-scale studies have increasingly found phthalates, bisphenols, perfluoroalkyl substances, flame retardants, and certain pesticides to have associations with obesity in humans. Had the National Children's Study continued as planned, we might have a much greater awareness of chemical exposures now, perhaps even placing these exposures on the typical differential diagnosis or workup for obesity and diabetes.

Back in 2009, during my first visit with Michelle's family, I was struck by the ongoing upsurge in her weight out of proportion to her height, which had jumped from the overweight range to above the 95th percentile, the threshold for calling a child obese. Just to be sure we couldn't make any additional improvements in her diet, we referred her to a nutritionist and set up a system so that she could earn a reward if she avoided sugary snacks for 2 weeks. I wanted to understand Michelle's situation in more detail, so I stepped up the frequency of her visits from every year to every 3 months. That way, I could check her weight more regularly and provide the necessary coaching and support to help her achieve her goals. She was clearly motivated to maintain a healthy diet, and we left that visit confident that she would see results even if it did not change her weight immediately.

Over the next couple of years, Michelle made amazing progress academically, even going to a national chess tournament. However, at her 9-year well-child visit, she reported being thirsty and urinating frequently. I also noticed patches of darker pigmented and thickened skin on her neck, which worried me. This condition is known as acanthosis and can be an early warning sign of prediabetes or diabetes. So I asked the family to come back for a fasting blood draw to check Michelle's blood sugar and insulin, as well as her lipids. Unfortunately, her laboratory tests confirmed my worst fears. We referred her to an endocrinologist for management of type 2 diabetes.

The focus at the endocrinologist's office was on managing blood sugars, but I wondered if chemical exposures *in utero* might have triggered Michelle's diabetes. Or was it simply preordained by the growth restriction in pregnancy? Could we have done something to limit chemical exposures earlier to prevent her developing diabetes? Over the next few years, Michelle's diabetes stayed under control, much to everyone's relief.

I wouldn't say she was so lucky, though. You might also wonder why I've included Michelle as an example of EDC dangers—after all, her situation might not seem so dire. However, that's my point: her tendency toward obesity and the development of type 2 diabetes may have occurred due to her exposure to chemicals and despite her healthy lifestyle!

———

Between 2002 and 2012, the rate of newly diagnosed cases of type 1 diabetes increased about 1.8% per year, and type 2 diabetes rates grew at an even faster rate, 4.8%. That pace would result in a doubling in the percentage of diabetic children by 2026.[148] Among 10- to 19-year-olds, rates for newly diagnosed cases of type 2 diabetes grew fastest among Hispanics and females, placing Michelle right in the center of this epidemic. A recent review of the scientific literature by Robert Sargis and colleagues at the University of Chicago describes how Latinos, African Americans, and low-income individuals have much higher exposure to so-called diabetogens, or chemicals that may induce diabetes. These include the PCBs and dioxins that were banned in the 1970s, certain pesticides, multiple chemical components of air pollution, bisphenol A, and phthalates.[149]

It's no use to look at the genome to explain the disparities or the increases in obesity and diabetes. It is unlikely that the human genome has changed significantly in a single generation to explain the increased susceptibility to excess weight gain in early life. So how else do we explain these twin epidemics? We're left looking for an environmental factor, or perhaps many environmental factors, to explain the epidemic we are seeing today. Indeed, the increases in obesity and diabetes are not restricted to the United States. Not every country measures these rates as carefully as the United States does, but the data we have suggest a worldwide increase concentrated in developing countries.

Let's start by making a visit to the laboratory of Bruce Blumberg at the University of California, Irvine. He and Jerry Heindel, a now-retired scientist at the National Institute of Environmental Health Sciences, helped spark my research into studying chemicals as obesogens in kids and my focus on putting a human face and some numbers to the endocrine disruptor problem.

OBESOGENS

Tributyltin is not a household name unless you are a shipbuilder or avid seafarer. Since the 1960s, TBT, as it's called, had been used as a biocide on the hulls of ships to inhibit growth of barnacles, algae, and marine organisms that affect a vessel's performance and durability. "Blistering barnacles!" I can hear my older son say, quoting Captain Haddock from the stories of *Tintin*. Many years ago, copper was used on the bottoms of ships for this purpose, but nailing copper sheaths is more work-intensive than painting a hull, and when scientists found that TBT could fit the bill, many shipbuilders were relieved. Since then, TBT has been banned because of its destructive effects on marine life. TBT continues to be used to stabilize plastics used in food packaging. As it turns out, Bruce and his colleagues realized that TBT was a prototypical chemical obesogen, with effects that persisted three generations later.

TBT selectively activates a type of receptor on cells called peroxisome proliferator activated receptors, or PPARs. PPARs attach to specific parts of the DNA sequence to activate genes just downstream of them.[150] The pharmaceutical industry has long focused on PPARs, and they are a key target for medications intended to treat diabetes. The difference between the effects of synthetic chemicals like TBT and diabetes medications such as rosiglitazone is the types of cells that they trigger.[151] As a personal aside, I haven't told you that my career almost took me into the world of synthetic organic chemistry. As a junior working in a chem lab at Harvard, I worked on a project to design reactions that would produce drugs similar to rosiglitazone and treat diabetes by affecting this set of receptors. I tinkered with the idea of getting a PhD rather than an MD or pursuing both degrees. So, talk about coming full circle—I am clearly drawn to investigating chemicals!

Phthalates are particularly effective at interacting with PPARs. They are made up of esters of phthalic acid and can be classified into two categories: low–molecular weight (LMW) and high–molecular weight (HMW) phthalates. LMW phthalates are frequently added to shampoos, cosmetics, lotions, and other personal care products to preserve scent. HMW phthalates are used to produce vinyl plastics for flooring, food wraps, and intravenous tubing. Within the HMW group, di-2-ethylhexylphthalate (DEHP) is of particular interest. DEHP is used in industrial food production and in manufacturing a wide variety of consumer items, including hospital equipment, food wraps, and containers, because it makes certain plastics such as polyvinyl chloride (or PVC) soft and malleable. Think water bottles, fast-food packaging, and hospital IV drips. Unfortunately, plasticizers can migrate within the material and leach out over time, entering the environment and, often, the human body.[152, 153]

When PPARs are triggered by phthalates, they make the body respond to an influx of calories differently than it otherwise would. Normally, our liver keeps some sugar ready to be quickly broken down for spare energy in a form called glycogen. Let's say your glycogen stores are in good shape and you eat a healthy meal, following the US Department of Agriculture's dietary guidelines, with a good amount of protein that the body would normally process into the raw materials for muscle mass (with exercise helping, of course). Phthalates and other PPAR-active chemicals disrupt metabolic function, causing mismanagement of calorie processing and diverting this same meal into fat creation rather than muscle. While this phenomenon is much more complex in practice, the principle is simple: phthalates tell the body to grow fat cells when that might not be an optimal use of the incoming nutrition.

Phthalates can also cause inflammation and create an imbalance in the body called oxidative stress.[154] The pancreas is particularly susceptible to the effects of this stress, and insulin-related activities can go awry. Inflammation can also narrow arteries, contributing to heart disease.[155] Animal studies have raised another red flag when it comes to the heart: DEHP exposure can induce arrhythmia and cause dysfunction in heart muscle cells.[156]

Phthalates also have a negative influence on testosterone.[157,158] While DEHP can act on the arteries and heart muscle directly, the effects of low-molecular-weight phthalates on the male sex hormone also have implications for heart disease. In older men, reductions in testosterone have been associated with early death from heart disease.[159,160,161] Supplementing so-called low-T levels with testosterone would seem like an easy solution to deficiencies that may be EDC related or otherwise. Yet studies of supplementation have not consistently shown the expected benefit, and in fact some have suggested negative consequences.[162,163,164,165] Some have theorized that the decreases in testosterone might be markers of the diseases that cause cardiovascular problems instead.[166] There could also be an explanation here like the one we talked about for thyroid supplementation in pregnant women with subclinical hypothyroidism—it may be that the dose used to replace the deficiency was not sufficient or too much, depending on the person, limiting the success of the intervention. Mother Nature is tough to mimic. It may be more effective to work on preventing low testosterone than treating it.

The EDC Disease Burden Working Group brought together leaders in the field of obesogens and metabolic risks, led by Juliette Legler (now at Utrecht University), to review the latest science on EDCs, obesity, and diabetes.[167] This group of researchers noted associations between prenatal exposure to phthalates and child obesity but found the studies to have two major challenges. One is that boys likely differ from girls in their response to phthalate exposures. Another challenge relates to the Barker hypothesis I described at the beginning of the chapter. None of the studies were able to capture fetal growth. They were able to measure birth weight, but if these chemicals induce a "thrifty phenotype," without data from ultrasound measurements, you might miss the effect of *in utero* exposure. Another problem is that BMI is a somewhat crude marker of obesity (think about football players who have "obese" BMIs and you get the idea). Fat mass, as measured using body composition measurement, is ultimately a better marker of obesity than BMI, but the studies did not capture that data.[168]

The experts saw stronger evidence for phthalates triggering obesity in adults in the exquisitely executed and well-known Nurses' Health Study. For decades, researchers at the Harvard T.H. Chan School of Public Health have dutifully kept in touch with thousands of nurses from throughout the country and are now following their children. This population has told us much that we know about prevention today, including the benefits of quitting smoking and reducing alcohol consumption to prevent colon cancer and the benefits of a so-called Mediterranean diet on heart disease risk.

In the next chapter, you'll come to know Dr. Russ Hauser's studies documenting phthalates and their effects on reproductive function. Here I'll discuss how his work influenced the thinking of obesity and diabetes researchers. His group at Harvard analyzed data and specimens from the Nurses' Health Study and found a striking pattern: 10 years after measuring exposure to phthalates, the higher the women's phthalate exposure, the more the weight gain.[169] Two other research groups conducted similar studies, but they had only followed subjects for 1 to 2 years, hardly enough time to see a pattern that was convincing (though one study of seniors from Sweden found weight gain even in that time frame).[170] Together with the very strong laboratory evidence, the probability of causation was still judged to be roughly in the range of a coin flip, in part because the EDC Disease Burden Working Group wanted to see more and better human studies before becoming more strongly convinced.

Earlier, I explained how our progress in understanding obesogens has been hamstrung in part by the cancellation of the National Children's Study. To compensate for that, the scientific community continues to comb through available data from other studies, including the Centers for Disease Control and Prevention's surveys. One limitation of these studies is that they measure exposure and outcome at the same time. In describing criteria for causality, Sir Austin Bradford Hill emphasized that causality could be determined only on the basis of studies that documented effects following exposure and not on those that simply measured exposure contemporaneously with the disease.[171] Another limitation in interpreting studies that measure obesity at the same time as chemical exposures is that phthalates are fat soluble. Obese children could have higher levels of phthalates in their urine because it commonly accumulates in fat cells, explaining an association between high phthalate levels observed in urine and childhood obesity.[172,173] This is called reverse causation. When interpreted with the right amount of caution, I have found these data very useful for generating hypotheses about obesogens, although I have been very careful not to imply anything definitive.

I am more comfortable, however, interpreting measures of insulin resistance from the CDC's data and relationships with chemical measurements at the same time. That's because pancreatic function and blood pressure are much more responsive to stressors than body mass, which takes time and continuous effort to change. Reverse causation is also less of an issue here. A series of studies I have conducted with a terrific research scientist on my team at NYU, Dr. Teresa Attina, have suggested that phthalates can influence metabolism and cardiovascular risks in kids. We first looked at data from 2003 through 2008, when DEHP was more widely used, and found that fasting measures of insulin resistance in adolescence, measured in blood, were higher in direct relationship to the levels of DEHP breakdown products in the urine.[174] The same was true for blood pressure,[175] with the effects occurring in kids as young as 6. The CDC doesn't draw blood in subjects this young, let alone ask young kids not to eat for 6 hours or more, so we don't have data on fasting blood sugar levels. Then we looked at newer data, from 2009 through 2012. Here we didn't see the association with DEHP so strongly, but different phthalates known as replacements for DEHP, called DIDP and DINP, showed the same patterns with higher blood pressure and more insulin resistance.[176,177]

IS AIR POLLUTION AN OBESOGEN?

Air pollution is getting increasing attention from global leaders, especially because air pollution and climate change are closely related. Fossil fuel burning and other industrial activities that emit carbon dioxide also emit other chemicals we inhale that worsen asthma in affected children. Taking a closer look at air pollution, it's important to consider how it's made up of a complex mixture of gases and particles that are smaller than the naked eye and very difficult to detect. The small size of the particles allows easy entry into the bloodstream, where they can inflame the coronary arteries and contribute to heart attacks, causing the same kind of difficulty in the brain's blood vessels that can lead to strokes. Researchers at the University of Washington who lead a Global Burden of Disease program to rank risk factors for disease put outdoor air pollution near the very top, killing 2.5 million people each year worldwide.[178]

The Global Burden of Disease researchers don't yet measure endocrine disruptors and their effects, but air pollution's impacts included endocrine disruption, because of the chemicals in the particles. Metals like nickel, cadmium, and mercury are big components as well, and those metals may also disrupt hormones. In addition, the burning of fossil fuels produces polycyclic aromatic hydrocarbons (PAHs) that also have a variety of effects. Some are estrogens,[179] while others antagonize male hormone functions.[180] PAHs can disrupt thyroid hormone.[181] They can also affect the PPAR receptors that play a crucial role in lipid and sugar metabolism.[182] Ginny Rauh's colleagues at Columbia have identified increased risk of obesity based on body mass index measurements and increases in body fat at age 7 in children born to mothers with higher exposure during pregnancy.[183]

Air pollution is also known to cause inflammation and oxidative stress in the body, especially in the pancreas, which is particularly susceptible to damage from these processes. As it turns out, Dr. Kathrin Wolf and colleagues from the German Research Center for Environmental Health found that adults with higher particulate matter exposure had more insulin resistance, especially among those with prediabetes.[184] They also found higher levels of leptin, a hormone used to communicate metabolic signals across the body. There are other human studies that are increasingly suggesting air pollution as a cause of diabetes.[185] Yet, when it comes to mapping out a plan for combatting the growing global pandemic of diabetes, air pollution still does not get the needed focus. When scientists speak about the environment and diabetes, they are emphasizing the physical and built environment.[186] This myopic approach fails to embrace the strong scientific evidence implicating the chemical environment.

BISPHENOLS AND OBESITY

Like the PBDEs and phthalates associated with brain disorders and DES and DDT with other abnormalities, BPA is another chemical for which we may be suffering the consequences of failing to act when science sent early warning signals. Laboratory studies of BPA have shown that it has many of the prototypical characteristics of an obesogen: It makes fat cells bigger and is a synthetic estrogen like DES, though less active. It also counteracts the function of adiponectin, a hormone that protects against heart disease.[187,188,189]

Bisphenols are used in metal food and beverage containers to prevent corrosion. What manufacturers did not know at the time these containers were developed was how it degrades and leaches from can coatings and into the food they contain.[190] In order to better understand how kids are exposed and the most dangerous sources of BPA exposure, one revealing study of preschool children collected dust and indoor and outdoor air as well as food samples. The findings confirmed that 99% of BPA exposure comes through solid and liquid food.[191] All foods are susceptible to contamination, and acidity doesn't make this worse; some studies suggest neutral pH foods have the highest levels.[192]

In 2017, Juliette Legler and I published in the journal *Environmental Health Perspectives* a comprehensive and systematic review of the literature on early-life exposure to BPA and its effects on obesity-related outcomes in rodents.[193] We also performed a meta-analysis, a method used to integrate the results from multiple studies. Across 61 studies we found strong increases in body weight, fat weight, and free fatty acids, as well as an intriguing trend toward possible increases in leptin (but not significant enough to dismiss the possibility of being explained by chance). The findings for leptin were important because it is a hormone that regulates metabolism and can predict patterns of growth in children in the first 3 years of life.

In Ginny Rauh's cohort of children from northern Manhattan and the South Bronx, prenatal BPA exposure was associated with fat mass and percent body fat in children at age 7.[194] In a Spanish study, higher maternal BPA levels during pregnancy were found to be associated with higher BMI in children by age 4—like Michelle.[195] A California study of Mexican American mothers did not show the same pattern, revealing increases in BMI in 9-year-olds in association with BPA levels in the children rather than during pregnancy.[196] An Ohio study examined an African-American birth cohort, with the children demonstrating a more rapid increase in BMI between ages 2 and 5 among infants with higher BPA levels.[197]

These studies are not as consistent as the laboratory data, though differences in results from humans could have a few explanations. BPA generally exits the body in a few days in most cases,[198] and these studies generally examined one or at most two urine samples across the many months of pregnancy. Also, remember the Barker hypothesis from the beginning of the chapter? A Dutch study found higher urine levels of BPA in pregnancy associated with slower fetal growth,[199] just the kind of "thrifty phenotype" pattern Michelle experienced. More work with this population is underway to see if the prenatal BPA exposure is associated with the rapid postnatal weight gain Michelle could not fully reverse.

Back when Bruce, Juliette, and the other members of the EDC Disease Burden Working Group considered BPA effects on childhood obesity, the BPA findings from Ginny's cohort were not available. This is important because the Columbia study measured fat mass in children in relationship with prenatal BPA exposure.[200] Had the panel had such information, we might have rated probability of causation higher, but we can't Monday-morning quarterback these decisions. The panel could not fully agree on a narrow range for the probability of causation, because the human evidence was not strong enough. Some were more willing to accept a roughly one-half probability of causation, while others faulted the design of the human studies as a reason for missing important findings and rated the probability more like a 1-in-3 chance. If the research confirms BPA is a risk factor for childhood obesity, the consequences of exposure to BPA are substantial both in the United States and Europe, with 33,000 additional obese 4-year-olds in the United States and roughly 42,000 more in Europe.

What does this boil down to?

BPA exposure may explain nearly 2% of all obesity in 4-year-olds. That may not seem like much to some, but it's important to emphasize the many different causes of obesity—genetics, diet, physical activity, "built environment," and other chemicals also contribute. A bigger percentage, say 10%, would actually be difficult to believe. What's not hard to believe is how the costs of such a small percentage translate to the economy in a big way. Just like adults, children who are obese have more health care costs than normal-weight kids. Children who are obese are more likely to be obese adults. This means that health care costs of those obese adults—as well as lost years of health—can be pinned directly onto obesity in childhood. In the United States and Europe, the costs of BPA exposure in 2010 were $2.4 billion and $2.0 billion, respectively.

Remember what I also said about BPA and the way it can block the heart-protective hormone adiponectin? David Melzer and his colleagues at Exeter University in the United Kingdom have published a series of studies in humans corroborating concerns first identified in the laboratory.[201,202,203] The studies vary in their design, and the populations are different (one from the United States and two from different cities in the United Kingdom). The results were striking, finding increases in coronary artery disease, whether measured by diagnosis or injecting a dye into the coronary arteries to measure the degree of narrowing. Recently, I estimated that nearly 34,000 new cases of coronary heart disease resulted from BPA exposure in the US alone, with costs of over $1.7 billion annually, based on continued exposures.[204]

The costs of these diseases are very high. Shouldn't we be looking at ways to mitigate them? If avoiding chemical contamination increases health and lowers unnecessary medical expenses, isn't that a win-win?

WHAT YOU CAN DO NOW

Some of the obesity-promoting effects and cardiovascular risks of these exposures are, thankfully, reversible. Choosing personal care products labeled "phthalate-free" reduced urinary levels of one LMW phthalate by 27% in young girls in one study.[205] Food packaging is a major source of phthalate exposure.[206] The more we all eat fresh foods, the less exposure there is to DEHP, which can negatively affect proper lipid and sugar metabolism.[207] Consumption of a Mediterranean diet has also been associated with lower levels of PFOA.[208,209] We typically describe these diets as "healthy" because they are more predominantly populated with leafy greens and vegetables that protect the heart because they are heavy in antioxidants. These diets also are less likely to be contaminated with persistent organic pollutants that may damage the endocrine and cardiovascular systems.

Another step you should take is to get to know your plastics. Plastic containers are supposed to have a recycling number on the bottom. The number 3 is for phthalates, 6 is styrene, and 7 was previously BPA but is now categorized ambiguously as "Other" and may contain BPA replacements that are as problematic. Ever since the FDA banned its use from baby bottles and sippy cups, most companies that manufacture plastic bottles and containers have opted out of using BPA. However, BPA-free does not necessarily mean bisphenol-free.

Never wash any plastic in the dishwasher or put it in the microwave. The heat degrades the plastic and releases the chemicals into the food, water, or other substances. If it's meant for single use, keep it that way and only use it once. As soon as you notice that plastic has been "etched," stop using it and recycle it safely. Note that using glass containers avoids this problem altogether. In our household, we try to buy food in glass containers rather than plastic whenever possible, and only store leftovers in glass containers. Reducing canned food consumption is another way of reducing bisphenol exposure.

If you don't need that inkless thermal paper receipt, don't take it. Increasingly, companies are sending email receipts instead, keeping a more permanent record for documentation purposes.

THE WILD WORLD OF -OMICS AND HOW THEY ALSO CAN HELP

More good news may be on the way. Earlier in the book, we discussed the influences of chemical exposures on the expression of certain genes. Dana Dolinoy and her colleagues in Michigan conducted a study with mice that showed how dietary exposures can modify the effects of chemicals on genes. Folic acid consumption appeared to limit the effects of BPA exposure on obesity in mice. These so-called agouti mice (named for their genetic abnormality that can produce a vivid yellow color) showed the promising ability of a dietary intervention to reprogram the genome to normal. Agouti mice born to unexposed mothers had a darker color, while increasing levels of BPA shifted their color toward yellow. The researchers measured agouti gene expression in the mice, and found it also increased in direct relationship to the BPA exposure. Supplementing the mothers of BPA-exposed mice with folate both changed their color back to brown and reduced the expression of the agouti gene.[210]

Are we there yet with humans? Such prescriptions are still a decade or more away, but speak to the potential of epigenomics, a relatively new science that looks at the consequences of changes in gene expression that can occur without modification of the code itself, to jump-start efforts at prevention.

The explosion of *-omics* (a suffix increasingly used to denote several fields of biology: the study of genetics, known as genomics; proteins, known as proteomics; small molecule breakdown products of metabolism, known as metabolomics, among others) has opened other fields of opportunity as well. One of the challenges of obesity research is that we don't have many barometers of metabolic health that can provide early warning signs and opportunities to reverse damage before it is permanent. There is an entire field of metabolomics screening and measuring hundreds of chemicals in the human body that are produced in small amounts by enzymes and other parts of our molecular machinery as they process diet and other environmental inputs.[211] Imagine a world in which we have early warning systems to detect metabolic disruption by synthetic chemicals. This would take two steps: (1) finding a pattern of metabolomic changes that chemicals like BPA induce and (2) documenting the impact of these metabolomic changes. Studies are also examining metabolomic changes as they may be induced by diet and other environmental factors. Perhaps this won't help Michelle, but it may help Michelle's children as they grow.

By replacing bisphenols in can linings with a safer alternative free of obesogenic effects, we can avoid thousands of cases of childhood obesity and tens of thousands of new incidents of coronary heart disease. One replacement candidate is oleoresin, though additional testing is needed to confirm its safety. Consider the economic gains and obvious incentive for replacing bisphenols: The potential cost of one BPA alternative, oleoresin, is about 2.2 cents per can. With annual production of 100 billion food and beverage cans, the cost of replacement would be $2.2 billion. When compared with the medical costs for childhood obesity and adult coronary heart diseases, the results are clear. Using 2008 figures, BPA exposure was estimated to be associated with 12,404 cases of childhood obesity and 33,863 cases of newly incident coronary heart disease in the US, with estimated social costs of $2.98 billion—and that was 10 years ago!

Removing BPA from anything to do with food might prevent 6,236 of the 12,404 cases of childhood obesity due to BPA exposure, and 22,350 of the 33,863 cases of newly incident coronary heart disease due to BPA per year. Those may seem like small numbers, but the potential annual economic benefits of each case prevented are so large that they total $1.74 billion.[212] Although more data are needed, these potentially large health and economic benefits could outweigh the additional costs of using a safer substitute for BPA. These benefits are unlikely to occur if BPA is simply replaced with BPS, BPP, BPF, BPZ, BPAP, or other bisphenols. What little we know is that BPS is as estrogenic if not more so than BPA, and more persistent in the environment and equally toxic to embryos.[213,214,215,216,217,218,219]

In today's health-conscious society, many of us, myself included, do all that we can to eat right, stay in shape, and live healthy lifestyles. Yet many people who are stricken by type 2 diabetes and obesity arrive at those conditions through no fault of their own. Like Michelle, they may eat a healthy diet, exercise regularly, and still fight excessive weight gain. We can do a lot to protect everyone from "lifestyle" diseases that are caused by things other than lifestyle.

CHAPTER FIVE

A REAL-LIFE *CHILDREN OF MEN*?

In 1992, P. D. James published a novel describing a future in which sperm counts plummeted to zero. Called *The Children of Men*, the story is set in England in 2021, where chaos has resulted from the growing population crisis. The Baroness James of Holland Park, as the author is also known, could be faulted for exaggerating the timing of the decline in sperm count in her now-famous novel. Unfortunately, that book and its subject—decreasing sperm counts—was a harbinger of a much larger hidden crisis still to come.

It's perhaps no coincidence that the same year *Children of Men* was published, a Danish pediatrician, Niels Skakkebæk, reported findings in the *British Medical Journal* substantiating P. D. James's concern that male fertility was in decline. Pooling data from 14,947 men from 61 different studies worldwide, he documented decreases in semen quality between 1938 and 1991.[220] This was not the only problem afflicting the male reproductive tract with increasing frequency. In 2001, he coined the term testicular dysgenesis syndrome, or TDS, to describe a series of conditions in males sharing a common origin.[221] Very early in life, the embryo is programmed in a specific way to set in motion the steps that produce the male gonads. If this programming is disrupted, multiple consequences can occur at different points in life, but the consequences are not the same in every case. The first and most obvious flag of TDS is a misplacement of the urethral opening at birth, called hypospadias. Rather than at the head of the penis, the urethra emerges at the base or farther down the shaft, even near the scrotum. The foreskin is frequently used as part of the repair for this condition, making it impossible for affected boys to be circumcised immediately after birth.

Another sign of TDS that can be noted at birth is the failure of the testicle to descend fully, a condition known as cryptorchidism. Sometimes the testis will descend later in infancy, before surgical repair is indicated. A major reason for intervening when the testis does not descend is to reduce the risk of another condition associated with TDS: testicular cancer. The relationships among these conditions are complex and variable: not all boys with undescended testis will develop testicular cancer, and not all boys with testicular cancer have had undescended testis or hypospadias.

MALE REPRODUCTIVE DISRUPTION

While it's been difficult for scientists to track or gather trend data (statistics that would show changes in rates of conditions over time) for undescended testis, the data for epidemic increases of hypospadias and testicular cancer are particularly alarming. Denmark shows a doubling in hypospadias between 1977 and 2005.[222] In the United States, the Centers for Disease Control and Prevention documented the same doubling in the 1970s and 1980s, using data from the US Birth Defects Monitoring Program.[223] The testicular cancer trend data is even more consistent across countries with the data, led by Denmark, Norway, and Sweden but also the Czech Republic, Bulgaria, Spain, Austria, the Netherlands, Poland, Finland, Estonia, Lithuania, and Latvia, just to name a few countries.[224] In the United States, between 1971 and 2004, testicular cancer increased in frequency by 71%.[225]

Niels uses the term TDS to speak to the reality that disruption of the male genital tract is much more frequent than previously thought and seems to be increasing. While genetic causes do likely exist, the heritability of testicular cancer ranges from 37% to 49%, suggesting that genetics does not explain the majority of TDS.[226] Genetics also can't explain the trends; DNA doesn't change that quickly over a generation. Studies of dietary factors were first conducted in 1975, but initial studies suggesting fat intake as a contributor have not been corroborated with newer, better-designed studies, leaving the role of diet as inconclusive at best. Researchers have also looked at physical activity but have not seen a consistent signal in relationship to testicular cancer.[227]

That leaves environmental factors as the leading suspects. Research has proven that chemicals that act as synthetic estrogens (BPA and DES, for example) or blunt the effect of testosterone and other androgens can induce hypospadias, undescended testis, testicular cancers, and low sperm counts. We also know that an insulin-like hormone is expressed by a testicular cell type called the Leydig cell. This hormone influences the growth and differentiation of a key embryonic structure that orchestrates the descent of the testis. Knock out the gene for this hormone, and you will have mice with both testes failing to descend. We know that this gene is regulated by estrogen. Put all of these pieces of the puzzle together and you have probable cause that synthetic chemicals can contribute to all three TDS conditions by disrupting hormones.[228]

A patient I encountered during medical school illustrates the challenges that make the individual impacts more ambiguous, even though they are real on a broad, population basis.

ESTEBAN AND THE MYSTERY OF TESTICULAR CANCER

I met Esteban during my first medical school rotation in July 1996, just after I passed the first of the three examinations an aspiring doctor needs to conquer before applying for a full medical license. As medical students watched and supported procedures in the operating room, senior attending surgeons would stop trainees at random moments to quiz them for arcane information. *Sabiston Textbook of Surgery*, in its 14th edition then, was a bit too bulky to carry around, so in spare moments between surgeries I would pull out my 8-megabyte HP 200LX Pocket PC to study acronyms and mnemonics. Medical students these days can rely on the e-book version of the 20th edition, though the electronic files would have been too big for the then-state-of-the-art Pocket PC to handle!

Esteban's parents, who were farmworkers, had immigrated to the United States just before his mother became pregnant with him. He had moved to the East Coast to attend college, and we bonded over the fact that we were both the first in our families to make such a leap. My parents had immigrated to the United States a few months before I was born, and like Esteban's, my working-class parents had worked hard to pave the clearest path for their children to excel. I still remember doing my homework in offices my mother cleaned late at night, navigating the near-empty subway to eventually get to our rent-controlled apartment in the Village and to bed.

Esteban was just about to graduate from college and was looking forward to attending law school when he felt a lump in his scrotum. The internist at the university's health service confirmed the lump and immediately referred him to the urology clinic where I was rotating. It was testicular cancer.

Esteban's family had no prior history of testicular issues, but Esteban recalled a concern that one of his testicles had not fully descended until later than expected, though it didn't require surgery. Undescended testis has long been known to indicate a higher risk for testicular cancer. Back then, the *Sabiston Textbook of Surgery* explained how surgical intervention to place the testis in the scrotum, called orchiopexy, was indicated for children and adolescents with cryptorchidism. At the time, the troubling trends in testis cancer and sperm count had not placed endocrine disruptors on the conventional medical radar, and Niels had not yet published the paper coining the term TDS. So our focus in the clinic was on management, not on identifying a cause.

The farmworker history of the parents did not register to me then. The typical medical school curriculum gives extremely short shrift to occupational exposures, and I cannot tell you for certain that Esteban's parents were exposed to pesticides. But let's take it as a given that farmworkers in the late 1970s working in the fields of California were almost certainly exposed to pesticides on a daily basis. So-called take-home exposures are quite common for other family members who are not themselves working around pesticides. If workers do not change footwear or clothing, they carry home residual dusts that contain smaller amounts of the pesticides used in the fields.[229]

We didn't begin to study the risk of testicular cancer to farmworkers until the 1980s, and since then, the results have been varied and inconsistent. More recent studies, however, look at the people who apply pesticides: clear increased risks have been found in Sweden, the United States, and the United Kingdom. All of the studies have relied on self-reported use of pesticides, which is a problem for many reasons. There are potential biases in recall among those who have testicular cancer, and there is no easy way to tease out which pesticides were used. In addition, it's nearly impossible to quantify the magnitude of exposure and compare groups in a way that everyone agrees is meaningful.[230] Indeed, the technology to measure minute amounts of these pesticides was just emerging as I met Esteban.

I was one of the last generation of medical students who were generally expected to work 24- to 36-hour shifts as part of our training. Forgive me if I am fuzzy about a few details of Esteban's story because of sleep deprivation that is fortunately a thing of the past.

Early studies that measured levels of pesticides in the body to evaluate DDT and other exposures did find effects, however. And these older studies done on DDT and its major metabolite DDE still hold true, showing consistent associations with testicular cancer.[231] Two studies that collected serum to measure exposure before the diagnosis, minimizing bias further, both showed associations with cancer.[232,233] Thankfully, DDT was banned in agricultural use in the United States in 1972, a few years before Esteban's birth, in large part due to Rachel Carson's efforts. While that might seem to exclude a possible role for DDT exposure in Esteban's case, his mother could have been exposed in years earlier and had higher levels of DDT in her serum during her pregnancy with Esteban.

There is also one intriguing, though as yet not independently confirmed, study linking flame retardants with testicular cancer. Lennart Hardell examined mothers of men with testis cancer in Sweden, finding greater levels of PBDEs in the mothers compared to a matched comparison group.[234] The problem with this study is the PBDE measurement occurring so far after the relevant exposure in pregnancy. PBDEs do have negative effects on androgens in the laboratory.[235] There is an intriguing literature on effects in firemen who might have greater use of flame-retardant material, but it's inconclusive.[236] The studies in firemen used questionnaires instead of measuring exposure using a serum sample. The inconsistency also speaks to the difficulty in studying a condition that fortunately occurs rarely. Though it is the most common type of cancerous malignancy in men 15 to 40 years of age, fewer than 6 men per 100,000 are diagnosed each year, according to data from the National Cancer Institute.[237]

PBDEs were widely used across the United States when Esteban's mother was pregnant with him; the wide use was due to the California law requiring flame retardants to be added to furniture. Would measuring PBDEs in his mother's blood at that time have detected a high level? I can't say. Could there have been some other factors explaining the association found in Sweden? Yes.

But back in Boston, Esteban had blood tests and imaging studies to evaluate the stage of the cancer. Though his ultrasound scan suggested the more aggressive type of cancer, most of the other studies brought good news. The tumor had not spread elsewhere in the body, except for two lymph nodes that seemed enlarged, and they were less than an inch in size. Only one of his tumor markers was slightly elevated, putting him in what today is classified as a stage IIA seminoma. Even with the spread to lymph nodes, the typical survival rate for a stage IIA seminoma is 96% after surgery and radiation. Regular blood test monitoring, effective chemotherapy, and radiation have transformed this disease into a treatable condition.[238]

Some of you might wonder why I am writing about a patient whose condition is treatable. Yes, we live in a society in which medical advances have transformed conditions that were once fatal and are now manageable. That's great. I went into a career in medicine in part because of the power of medical innovation. I also appreciate that it may take less time to write a prescription for a medical problem, and more effort to focus on the primary and preventable causes of conditions. There is also substantial irony in that chemical exposures appear to induce conditions that then require other chemical exposures (pharmaceuticals) to cure. It was no small challenge for a rising legal star to have to detour his professional growth, having travelled so much distance personally and otherwise to achieve so much, perhaps beyond what Esteban's parents ever dreamt. And so, even with my sleep deprivation, this story stuck with me as a failure of the medical profession to prevent diseases, as much as we have developed terrific techniques to treat many of them. It's also important to emphasize the effects of EDC exposures on people's lives, even when they are treatable. And treatment may have consequences of its own: chemotherapy is well known to increase risks for a secondary cancer later on.

PBDE EXPOSURE, TESTICULAR CANCER, AND CRYPTORCHIDISM

In 2007, a study of breast milk and placental tissue in mothers of Danish and Finnish boys with cryptorchidism, as well as a similar group without the condition, were used to assess exposure in the earliest phases of life. Consistent with Niels's TDS hypothesis, PBDE levels in breast milk were much higher in the mothers of boys with cryptorchidism compared to the others. There were not differences in placental levels of the flame retardants. However, PBDE is very fat soluble, meaning it accumulates much less in the placenta, making differences in this study harder to detect than for breast milk.

Cryptorchidism is more common than testicular cancer but still affects only 1 in 100 boys.[239] Based on the available data, we estimated one-fourth of the population to have higher risks of undescended testis, and 47% to 86% increases in the most highly exposed children. That translates to 4,300 boys each year needing surgical correction to prevent the development of testicular cancer. The experts rated the probability of cause-and-effect to be much higher for PBDEs with cryptorchidism (roughly a coin flip) compared to testicular cancer (more like 1 in 10).[240]

Three years after the expert panels published their work, a Canadian study in *Environmental Health Perspectives* found higher levels of PBDE in the hair of mothers who had just given birth to newborns with cryptorchidism than their peers.[241] And, as this book is being finalized, a new study from a different population in Canada found higher hair levels of PBDE among mothers of babies born with hypospadias.[242] Had these studies been available earlier, the EDC Disease Burden Working Group might have reached a stronger conclusion.

OUR DWINDLING FERTILITY

I met Martin in the newborn intensive care unit in late 2000, in my second year of residency. Martin was born 8 weeks premature to a couple who had tried for over a year to get pregnant. Back then, getting insurance companies to pay for in vitro fertilization (IVF) was not easy; thankfully, since that time 15 states have passed laws to require insurance companies to cover infertility diagnosis and treatment, although not all of those 15 require IVF coverage. Leaving aside for the moment the substantial costs of IVF, Martin's mother, Susan, had been on an emotional marathon, having navigated 4 weeks of bed rest while getting intravenous steroids to accelerate the development of Martin's lungs in anticipation of an early exit from the womb. She had crossed the finish line when Martin was born.

Martin's parents were unusual in that they were both in their early 30s, younger than many seeking IVF today. They both had extensive medical workups to be sure there wasn't a simple, treatable explanation for their difficulty conceiving. A smattering of articles in the scientific literature described male workers exposed to synthetic chemicals having problems with fertility in the 1970s and 1980s,[243,244] but even if the infertility workup had included those questions back then, neither Susan, an accountant, nor her partner, David, a lawyer, would have given remarkable answers. Neither had hobbies or other activities that could have introduced unique or significant chemical exposures of concern.

Nowadays, the story of Susan, David, and their son, Martin, seems much more common. Fertility rates in the United States are at their lowest level ever recorded—59.8 births per 1,000 women in 2016 compared to 122.9 per 1,000 women at the height of the baby boom in 1957.[245] Other developed countries follow a similar pattern.

Of course, not all of the increase in infertility or the use of assisted reproductive technology procedures such as IVF can be explained by environmental factors. Many factors contribute to this trend, such as older age at conception and desire for a smaller family size. Between 1970 and 2000, the average age for a mother at the time of first birth increased 3.5 years to nearly 25. In 2014, it was 26.3.[246] Changes in cultural norms, access to contraception, education, and employment are among the factors that also need to be weighed in piecing together the story. Obesity is also on the rise and is known to have its own effects on male and female reproductive functions.

MEASURING MALE INFERTILITY

Estimating infertility rates can be complex. Researchers at the Centers for Disease Control and Prevention have used a periodic survey called the National Survey of Family Growth to try to estimate these rates. Studies suggest that the methods used in these surveys underestimate infertility, and yet data from the 2006–2010 survey indicate that 12% of married women report having trouble getting pregnant or sustaining a pregnancy.[247] Not measured in the CDC survey, time to pregnancy (TTP) is a common metric where infertility is defined as needing more than 12 months to get pregnant. Using TTP, carefully measured in population-based studies that follow couples trying to get pregnant, rates of infertility have been estimated to range even higher, as high as 24% in France.[248]

Another way to track infertility is to count medical procedures used to treat infertility. Indeed, there are more data for assisted reproductive technology (ART) procedures, which the CDC has compiled since 1996, when 64,000 were reported. That number has risen steadily, with over 115,000 ART procedures in 2002 and nearly 170,000 in 2014.[249]

Birth rates in many countries are hovering at or below the replacement rate, the rate of fertility needed to maintain the population. Denmark has perceived the ongoing decrease in fertility as such an urgent and real threat that its government developed an entire media and advertising campaign encouraging young couples to have children.[250] But what if the problem is not a lack of desire to conceive but a lack of capacity?

The film adaptation of *Children of Men* by Alfonso Cuarón is arguably better known than James's book, with three Academy Award nominations. In his adaptation, Cuarón changed P. D. James's story substantially, turning it into an issue of female infertility. As one of the most storied researchers in the field of EDCs and reproductive health, Germaine Buck Louis of George Mason University will tell you that any proper studies of fertility need to consider both partners' environmental exposures.

I focused earlier in the chapter on three conditions in the testicular dysgenesis syndrome—undescended testis, hypospadias, and testicular cancer. These all arise when development of the male genitourinary tract is disrupted. It's perhaps no surprise to imagine that if you can disrupt the plumbing of the tract, you can also disrupt the cells in the testicle that produce sperm. The most threatening potential consequences are the cancers that can arise in those cells whose development is disrupted. The testicle's ability to produce sperm is also limited. When Niels first wrote about TDS in 2001, decreased sperm count was the fourth condition in the bundle of male reproductive problems increasing in incidence, suggesting an environmental origin.

Indeed, a very recent study by Hagai Levine at the Braun School of Public Health in Israel and his colleagues replicated and extended Niels's analysis. Looking at data from 43,000 men in 185 studies conducted between 1973 and 2011, they documented a 59% decline in sperm count in Western countries. Men from Africa, South America, and Asia did not show a similar decline.[251] Several studies of men from infertility clinics have also found striking associations of phthalate exposure with decreased semen quality.[252,253,254] Though these findings have not been replicated in the general population of men,[255,256] it could be that men such as David coming to infertility clinics may be at higher risk of effects from phthalate exposure.

A difficulty with these studies is that they measure phthalate exposure in adulthood, well after the disruption of development, the proverbial car crash, has already happened. The driver, a prenatal environmental exposure, has already left the scene. For a detective trying to identify the driver that caused the accident, the problem is finding a fingerprint or other clue to solve the case. In the case of prenatal phthalate or other EDC exposures, that clue may come in a relatively unusual anatomic measurement, anogenital distance.

MEASURING AN UNUSUAL DISTANCE

Dr. Sheela Sathyanarayana is an associate professor of pediatrics and adjunct associate professor within the Department of Environmental and Occupational Health Sciences at the University of Washington. Sheela's decision to enter the study of endocrine disruption began with reading *Our Stolen Future* as an undergraduate, but the road that led to her career began with her mother. "My mom is an ob-gyn, and she cut out an article for me about an ongoing longitudinal study taking place in New York called the Study for Future Families," she told me.

This study was examining environmental causes of differences in male fertility and was led by Dr. Shanna Swan, who began her career as a statistician examining the effects of oral contraceptives. Through her work at the California Department of Health, she began to hear of broader concerns about environmental chemicals having similar and unintended effects. This drove her to eventually study the impact of plasticizing chemicals such as phthalates on genital development in infant boys and also on sex-specific behaviors in younger children.

Sheela and I first met through a program for pediatricians training in environmental medicine. We've worked together on various studies for more than 5 years and are now working with the four-city study called the Infant Development and the Environment Study, or TIDES for short. It is part of the Environmental Influences on Child Health Outcomes program I mentioned in chapter four. It's analyzing data from 50,000 children across the United States to understand the effects of EDCs and other environmental exposures.

Sheela's work has focused on the very specific effect of synthetic chemicals on sexual characteristics and development in males. She and her colleagues have already identified associations of first-trimester exposure to phthalates with increased incidents of genital abnormalities in male newborns in TIDES. One of the outcomes of Sheela's work is the realization, as she says, that "these chemicals never act alone; it's always a combination of factors, and it's therefore very difficult to appreciate their cause-and-effect complexity."

In her research, Sheela focuses on measuring anogenital distance (AGD). Measuring the distance from the anus to the base of the penis, the underside of the scrotum, or vagina, which is not part of the routine medical checkup, is important because it is a telltale marker of the relative exposure to male and female sex hormones. Boys have longer AGDs than girls because the cells in that portion of the body respond to testosterone exposure. More importantly, though, "this measurement is a harbinger for other consequences," Sheela explains. Shorter AGD in men has been associated with impaired male fertility, including lower sperm counts; longer AGD, a marker of greater testosterone exposure in early life, may have consequences for women. In animals, excess prenatal testosterone exposure can produce features of polycystic ovarian syndrome, or PCOS.[257,258] A recent study in China identified longer anogenital distance in women with PCOS.[259]

In animals, infant AGD correlates extremely well with later AGD, suggesting that this measurement does not typically change in response to other influences later in life.[260] In men, Jaime Mendiola and his colleagues have documented that men with shorter AGD have lower sperm counts.[261] Furthermore, animal studies document that low-molecular-weight phthalates such as dibutylphthalate, which are used in perfumes and nail polish, induce TDS and decreases in AGD.[262,263] Shanna Swan, along with Sheela and her colleagues in TIDES, documented decreases in AGD in association with first-trimester phthalate exposure in boys in a large, four-city US birth cohort.[264] A more recent Swedish study documented a similar decrease in boys in relationship to exposure to the DEHP replacement I talked about in the previous chapter, DINP.[265]

These two publications were not available when the EDC Disease Burden Working Group met in Copenhagen in 2014 to evaluate the effects of phthalates on male infertility. The Working Group relied on a landmark study by Germaine Buck Louis, the Longitudinal Investigation of Fertility and the Environment Study (LIFE Study), which carefully followed 505 couples planning a pregnancy. The study found an association between higher paternal urinary concentrations of two phthalates found in polyvinylchloride plastics, monobutyl and monobenzylphthalate, and longer time to pregnancy. The mother's exposures were not associated with time to pregnancy.[266]

Back then, Russ, Niels, and their colleagues rated the likelihood of a cause-and-effect relationship to be medium (roughly the proverbial coin flip). Would that assessment have increased with the new information from later studies? Possibly. There are always risks to making judgments, when new information can swing the pendulum in either direction. If you don't like the odds on a coin toss, the impact in the United States will make you more uncomfortable. The implications of exposure to the two phthalates Germaine and her colleagues associated with longer time to pregnancy translate to the need for tens of thousands of additional IVF treatments. The costs of the IVF treatment—not including the impact on children like Martin, who are born preterm, which is more likely in IVF pregnancies—is $8.8 billion.[267] This also does not include the costs to the mothers of IVF babies, such as lost work time for Susan due to the bed rest. Society also places substantial value on preventing emotional suffering such as that which David and Susan both had to soldier through.

WHAT DO ACETAMINOPHEN AND FOOD DYES HAVE IN COMMON WITH PHTHALATES? DO THEY ALSO PRODUCE MALE REPRODUCTIVE PROBLEMS AS SIDE EFFECTS?

Acetaminophen is well-known as the only pain medication obstetricians deem safe for pregnant women to take. But, like many drugs in pregnancy, while there have been studies of acetaminophen in animals, the safety of acetaminophen has not been fully assessed in pregnant women. When researchers started to recognize the effects of phthalates on the male reproductive tract, they also realized that acetaminophen and phthalates have a lot in common in their chemical structure. And, unfortunately, studies have begun to raise the alarm about acetaminophen and other analgesics commonly used in pregnancy.[268]

In animals, acetaminophen exposure shortens the anogenital distance in males. Human studies have identified higher rates of undescended testis in boys whose mothers took acetaminophen, particularly when exposure was in the second trimester.[269,270] Not all studies of pregnant women have confirmed this phenomenon,[271] but most of these studies have relied on questionnaires that ask mothers to recall their use of analgesics over a relatively long period in pregnancy, 6 to 8 weeks or so. Acetaminophen does not produce a long-term reduction in pain and fever because it is excreted from the body quickly, and so exposures in pregnancy are short and vary in their timing. That timing may matter for the male reproductive organs as they develop.

There's one other wild card. You may not have known that dyes used in food and clothing with a chemical structure called aniline can actually be converted in the human body to acetaminophen. Does that mean avoiding food dyes? We have a long, long way to go before I'd suggest that. Food dyes have created their own concerns; some have suggested eliminating them could reduce ADHD symptoms.[272]

Suffice it to say there are lower-hanging fruit in preventing endocrine disruption right now based on the available data, but it may add to the argument for consuming fewer highly processed foods.

The more compelling argument would be to reduce unnecessary analgesic consumption. Bernard Jégou and his colleagues at the Université de Rennes in France have shown ibuprofen to induce a syndrome like that found in older men in which their testosterone is clinically in the normal range but only because the hormone that stimulates production of testosterone (called luteinizing hormone) is revved up. The dysfunctional communication between the pituitary gland and testicle is called compensated hypogonadism. That seems innocent enough, but compensated hypogonadism is associated with reduced libido, reduced fertility, arthritis, cardiovascular disease, and diabetes.[273]

Martin spent about a month in the neonatal intensive care unit and needed minimal support for his breathing. Most of his care focused on monitoring his breathing to be sure that his brain was mature enough to maintain a healthy rhythm on its own and to build his feeding tolerance and weight gain without stressing his intestines, which were very delicate due to the lack of development in the womb.

While I lost touch with this family after I finished residency, the cognitive profiles of IVF babies in general are the same as naturally conceived babies.[274] Preterm babies, however, are known to have lower IQs compared to babies born to term. There also appear to be some differences in metabolic function and heart health, with slightly higher blood pressures and fasting glucose levels among children born from IVF pregnancies.[275]

The United States has one of the highest rates of preterm birth worldwide. You might wonder if environmental factors contribute to this important predictor of child health. The evidence suggests that they do: a study in the hospital where Martin was born, published by Kelly Ferguson (now at the National Institute of Environmental Health Sciences) and her colleagues, documented strong associations of a phthalate used in food packaging, DEHP, with higher odds of preterm birth, especially preterm birth precipitated by early labor or rupture of the membranes.[276] It's complicated to sort out whether phthalate-induced preterm birth is happening by an endocrine mechanism. Phthalates also induce inflammation and oxidative stress, which can make the placenta not function as well as it should.

EDCS, ED, AND YOUR PROSTATE

At the risk of making the men reading even more uncomfortable, I'd like to share some information on whether EDCs are contributing to erectile dysfunction. Testosterone is a crucial factor in male libido, and if phthalates and PBDEs work against testosterone function, then that possibility does arise. We know now that there is an association between phthalates and decreases in testosterone among men between the ages of 40 and 60.[277] Indeed, one study suggests that as many as 40% of 40-year-old men have some degree of erectile dysfunction,[278] and most men using medication for this condition are in their early to mid-50s.

The first report connecting chemicals and impotence dates to 1970, when a study in the *British Medical Journal* documented evidence that 4 of 5 farmworkers became impotent after using herbicides and pesticides. Their sexual function returned after ending contact with the chemicals and having a course of hormone therapy.[279] A more recent study in Canada did not find associations of plasma levels of persistent pesticides and

PCBs with erectile dysfunction,[280] though a Chinese study of workers in a plant using BPA for aluminum cans did associate high levels in urine with lower libido, erectile dysfunction, and difficulty with ejaculation.[281,282] Finally, in 2018, a study from Nigeria documented decreased levels of multiple reproductive hormones, including testosterone, among men working with electronic waste. These workers had higher levels of mercury, lead, cadmium, and arsenic. All these heavy metals are potential EDCs as well.[283]

I'm not a urologist, though this quick overview suggests this as an important area for further research. The more important and serious point to take away from this information about EDCs and male fertility is that these chemicals don't just affect women and children or men who are interested in conceiving. The World Health Organization and United Nations Environment Programme report from 2012 documents troubling increases in prostate cancer across the world as well.[284]

We know that the prostate is a hormone-dependent structure. Disruption of estrogen function, androgen receptors, enzymes that break down steroids, and even vitamin D have all been identified in the prostate. EDCs known to affect the prostate include pesticides, bisphenol A, PCBs, herbicides, and some heavy metals. Human studies have seen increases in prostate cancer in men exposed to certain pesticides, Agent Orange (the herbicide used in the Vietnam War), and arsenic.[285] This, too, is another area where we have a long way to go in understanding the effects of EDCs. As we'll talk about with breast cancer in the next chapter, human studies can take decades to conduct properly simply because the disease in question doesn't manifest clinically until 40, 50, or even 60 years following exposures.

CHAPTER SIX

THE CHEMICAL VULNERABILITY OF GIRLS AND WOMEN

Scientific attention to the chemical assault on women's bodies and long-term health is long overdue. Chemical exposures have been implicated in female conditions such as fibroids (noncancerous growths of the uterus that are painful and disabling and can reduce fertility), endometriosis (a condition in which tissue that normally lives inside the uterus is found outside the uterus and can cause pain and infertility), and even breast cancer, which can be fatal.

Just as Niels Skakkebæk described testicular dysgenesis syndrome, Germaine Buck Louis (who is now dean of the College of Health and Human Services at George Mason University, then at the National Institutes of Health) described a similar ovarian dysgenesis syndrome in 2007,[286,287] though the notion of ovarian dysgenesis is not new. Nearly 70 years earlier, Henry Turner of the University of Oklahoma described a syndrome in which girls born with a missing X chromosome are identified to have a slow growth rate with a series of characteristic features including a webbed neck (when a congenital skinfold runs along the sides of the neck down to the shoulders), broad chest, and widely spaced nipples. Women with what is now known as Turner's syndrome, also known as 45,X or 45,XO to refer to the missing X chromosome, are also infertile. Originally described as ovarian dwarfism, the syndrome was later described as ovarian dysgenesis. Ultrasonography of these women reveals obvious and clinical differences from women with 46 chromosomes. These include smaller ovaries with cysts in them or ovarian tissue that has no function. Fortunately, this genetic condition is rare, occurring in 1 out of every 2,000 to 2,500 female births.[288]

The current understanding of ovarian dysgenesis syndrome (ODS) is more subtle, but the underlying biology of the 45,XO and 46,XX forms is not so different. Perform an abdominal ultrasound on an adolescent with ODS, and unless she has already developed endometriosis or fibroids, the obstetrician or radiologist will not see any abnormalities. Perform a genetic test to count chromosomes, and you'll see all 46, including two X chromosomes. Yet look at the genes of these women with more careful epigenetic analyses, and they may have a variety of abnormal signals that herald the dysfunction. The disruption of the developing reproductive tract, including the ovary, fallopian tubes, and uterus, occurs *in utero;* but these internal changes don't affect the appearance of the girls when they are born. It's only later—decades later—that the evidence of cellular disruption becomes apparent, ultimately showing up in a woman as abnormal tissue in the vagina called *adenosis,* a smaller than normal cervix and vagina, uterine and tubal abnormalities, infertility, early menopause, and even breast cancer. Later in life these women experience difficulties with conception; pregnancy loss; and gynecological disorders such as endometriosis, fibroids, a lack of oocytes in the ovary that can divide and mature into eggs, or *ova,* that can be fertilized, and PCOS. Germaine identifies endocrine-disrupting chemicals as likely factors that produce these disabling, painful, and even fatal conditions when they interact with the genome.[289]

According to statistics, one or more of these conditions will affect 70% of women in their lifetimes. Already, 176 million women worldwide are diagnosed with endometriosis—a number more than likely underestimated because so many women go undiagnosed.[290,291] We've already discussed the substantial uptick in procedures for infertility in the previous chapter. It's important to add that endometriosis and fibroids are themselves important risk factors for infertility.[292]

In the previous chapter, I was able to give statistical evidence for increases in male reproductive conditions. It's likely that at least some part of the increase in assisted reproductive fertility procedures is due to decreases in female factor fertility, a group of conditions that ultimately affect a woman's ability to conceive and stay pregnant. The increasing use of donor oocytes (eggs) confirms this concern.[293] This procedure is applied to conceive a pregnancy when ovarian functioning is subpar or other aspects of the female reproductive tract are not working. The current data are not sufficient to parse the many factors, both male and female, driving these infertility trends. Numbers to confirm an increase in endometriosis and fibroids are not available. One factor that complicates our ability to study trends in these conditions is that procedures to manage these conditions have moved to the outpatient surgery setting, if they require surgery at all. Because there is less tracking of outpatient procedures, we have a big data gap.

If all of the female reproductive conditions that comprise ODS (endometriosis, fibroids, and female factor infertility) are increasing in epidemic proportions, why hasn't ODS made it to all the medical textbooks yet? Though I don't deliver babies or examine women as part of my clinical work, my colleagues in obstetrics, gynecology, and endocrinology have taught me a lot and have helped me understand why many women still don't know much about endocrine disruptors. Let's start with a story of a patient I met and referred to my endocrinologist colleagues.

EMILY'S EARLY PUBERTY

Emily was 6 years old when she was brought to the primary care clinic on a cold and snowy New York City Wednesday morning. Emily's father brought her to the clinic because he was concerned that she was developing breast buds. Emily's mother had begun puberty much later, at age 10, and so the timing of Emily's puberty was a bit of a surprise. Emily had been very healthy and had never been hospitalized for any condition. Her growth chart looked like the average US girl, as her height and weight reliably had tracked along the 50th percentile for her age. The resident took a history, performed an initial exam, and then presented Emily's story to me, one of the supervising doctors in clinic that day.

I asked the resident whether Emily was using any creams or lotions, thinking about potential sources of steroids that could be hormonally active and thereby induce breast development. I wasn't thinking about synthetic chemicals here. I was focused on pharmaceuticals with corticosteroids, which are typically used to reduce inflammation.

Why wasn't I thinking about EDCs? As much as the Endocrine Society has led efforts to increase general medical knowledge about endocrine disruptors, many endocrinologists today still don't regularly ask about them. Lack of knowledge regarding safe and simple steps to limit exposure may hold doctors back from even mentioning them. Some health professionals simply won't ask questions with answers that may expose knowledge gaps or lack of training, lead to awkward silence, or produce uncomfortable responses like "I don't know."

I asked Emily's dad about family medical history. When I came to cancer, he asked to speak with me privately for a moment. Apparently, Emily's mom had just passed away from breast cancer. Emily's dad hadn't wanted to bring it up because he was concerned about the stress that the family and Emily especially had experienced.

After the initial wave of sadness and empathy, I immediately began to think about the potential for a connection between breast cancer in Emily's mother and Emily's early puberty. Many breast cancers have cells with a higher proportion of estrogen receptors on their surfaces and grow in response to estrogen. Tamoxifen is a breast cancer drug that blunts estrogen effects — an intentionally endocrine-disrupting pharmaceutical.[294] Estrogen is a key hormone that stimulates development of the breast and the other female reproductive organs in puberty. Could synthetic estrogen exposures when Emily's mother was pregnant with Emily have set in motion the forces that produced both breast cancer in the mother and early puberty in Emily? Even if Emily's mother were alive and I had asked a detailed history about occupation, medication use, and personal care product and other environmental exposures, the answer I would have to give, based on the human studies, is typically uncomfortable for doctors to deliver: "I don't know." However, the animal and tissue studies raise red flags.

The resident and I presented a plan of action for next steps in Emily's care. We discussed a series of tests to rule out central precocious puberty, a condition in which the brain starts puberty too early by producing a hormone that accelerates the release of gonadotropins. Gonadotropins are hormones that stimulate growth of the gonads and tell the sex organs to make other hormones that begin sexual development. We were able to reassure Emily's father about this possibility in the end, as none of Emily's laboratory or other tests raised any red flags. We ended up defaulting to the notion of idiopathic precocious puberty. *Idiopathic* is a shorthand for not knowing the exact cause for a condition. At the time, even 4 years ago, this diagnosis seemed appropriate.

Today, reflecting on it, the conclusion we drew seems less sure. Could it be that chemical exposures in Emily's mother's pregnancy or even when Emily's mother was a child had induced a series of changes in hormone function that set in motion both Emily's mother's breast cancer and Emily's early puberty? Or could it be that exposures early in Emily's childhood had induced early puberty? We still don't have enough information to say either way. And as you read this chapter, you may feel like the evidence is more uncertain because it is. From an ethical or policy or personal perspective, left or right, how comfortable do we feel as a society if we find out 40 or 50 years from now that a chemical exposure could have effects across two generations or more? Remember that chemical exposures in a pregnant woman can affect at least three generations: the exposed mother, her child, and her children's children. Germ cells, the cells in the testis or ovary that contribute to a child, are programmed from the time the mother (or father) is in her (or his) mother's womb. Add in the complexity that signals for gene expression (without changes in the code) may get passed even farther down the family tree, and you can see the many ripples from the tsunami.

Could Emily's early puberty be an early warning for the breast cancer that stopped her mother's life short? Early puberty is a well-known risk for breast cancer,[295] but it's not clear how we can change later risk of breast cancer or the tempo of pubertal development. The ODS hypothesis suggests that the horse is out of the proverbial barn, though there is evidence suggesting that there are exposures that can contribute to accelerating or slowing puberty.

You may be surprised that I didn't raise environmental chemical exposures as an issue for the resident to address right away. I did raise this at the next primary care visit, after Emily visited the endocrinologist. There is no treatment that can undo chemical exposures, but there is so much we can do now to improve our lives going forward by reducing EDC exposures. Should doctors be communicating safe and simple steps to reduce exposure? The ever-shrinking length of the clinical visit will always present a challenge, with many other pressing issues and considerations competing with clinical care and especially direct treatment. Yet lack of time doesn't change the impact even a 60-second mention of steps to limit EDC exposures (as described in chapter 7) can have.

The fact also remains that puberty has sharply accelerated over the past 150 years. In 1850, the age of menarche (first period) was typically 15 in France; in 2000, the average age of menarche was 12.[296] The first reference ranges for genital development were developed by W. A. Marshall and J. M. Tanner in 1970.[297,298] We still use the term *Tanner staging* for girls and boys in primary care clinic to describe our evaluation, beginning typically at about 5 years of age. Based on their findings, the average age of breast bud development in girls was 11, and precocious puberty was defined as breast or pubic hair development before age 8.

In 1997 at the University of North Carolina, Marcia Herman-Giddens and her colleagues did a series of studies that set a new normal. Her data identified differences in the onset of puberty among girls of different races, with African American girls having more frequent evidence of precocious puberty.[299] There were some efforts to move the goalposts and reset the definitions, and a more recent study across three metropolitan areas in the United States confirmed substantial numbers of girls with breast development at ages 7 and 8. These findings are consistent with those from other parts of the developed world.

Is this new normal a good thing? Some have suggested that better nutrition in developed countries explains the shift to earlier onset of puberty. Others have blamed social stress as a factor that accelerates puberty. Obesity, which, as we know, has also increased in recent years, has been pointed to as another possible cause for this shift in age of puberty.

A large multisite study in the United States has begun to look at the possibility of chemical factors influencing puberty, as well as at other environmental exposures and genetics and their influence on breast cancer risk in young girls. So far, data from the study suggest that different chemical exposures can move along or delay pubertal development in girls. PBDEs were associated with delays in breast and pubic hair development,[300] while some phenols that are structurally similar to BPA and used in sunscreens (benzophenones) and antimicrobial soaps and toothpastes (triclosan) were associated with earlier and later breast development.[301]

These results appear inconsistent with each other, but again we come up against the reality that humans are not exposed to one chemical at a time. Chemicals can have different effects, depending upon the other exposures in the mix. For example, we used to think of chemicals as either acting like estrogen or acting like an estrogen inhibitor (and similarly acting like testosterone or acting like an antagonist to the male sex steroid system). However, with time, we've realized it's much more complicated. One chemical exposure may prime a receptor in the body to be even more sensitive to a second chemical than when the second chemical exposure occurs by itself. The same happens with pharmaceuticals—one drug may amp up an enzyme's activity and make a second more active. One chemical or pharmaceutical may slow or accelerate the metabolism of another. These effects can go both ways—there is antagonism as well as synergy.

One unavoidable limitation of the multicenter puberty study was that some of the chemicals can act like estrogens or disrupt testosterone function in a way that can be difficult to trace if you measure these chemicals in a single urine sample or perhaps two. Better studies are needed in which researchers more regularly measure these chemicals. Many investigators anticipate the large-scale ECHO study of 50,000 kids I described earlier will study prenatal and child exposures and impacts on puberty with repeated exposure measures.

The most compelling evidence that synthetic chemicals could be causing "the new normal" in girls and is part of a larger, overall impact on reproductive health comes from the DES story. We described very early in this book the landmark study by Arthur Herbst and his colleagues that documented the effects of fetal DES exposure on vaginal cancers in young girls. But that wasn't the entire story; DES-exposed women and mice have both been shown to develop fibroids.[302,303] DES exposure changes expression of genes and leads to disordered growth of uterine tissue that produces this condition.

We haven't talked about smoking much in this book, but studies of mothers who smoked during pregnancy tell us a lot here. Tobacco smoke is a complex mixture of chemicals, many of which are EDCs. Indeed, daughters born of mothers who smoke have more fertility issues. In mice, tobacco exposure has similar effects, reducing the number of oocytes and influencing sex steroid production in the ovary, potentially explaining the observations in humans.[304]

You've already heard a lot about BPA, so it should not surprise you that synthetic estrogens like DES and BPA can have similar effects on the female reproductive tract. In 2014 Jodi Flaws at the University of Illinois and her colleagues meticulously reviewed the literature, which told a much different story than a similar review completed only 7 years earlier. The change in the results speaks to the rapid growth of the field of EDCs and female reproductive health. Unfortunately, it describes BPA in stunningly similar terms to DES. BPA reduces the quality of oocytes in animals and women undergoing IVF.

It can impair the growth of the endometrium, the lining of the uterus, which needs to be rich in blood vessels for an embryo to implant and proceed to develop into a fetus. In humans, BPA exposure can induce excess androgen activity in part due to an overload of estrogen that is converted into testosterone elsewhere in the body. It can also induce sexual dysfunction and impair uterine implantation of embryos. The increase in androgen activity raises additional questions about potential involvement of BPA in PCOS.[305]

The evidence on BPA and its effects on the ovary was not available in 2014, when the EDC Disease Burden Working Group reviewed the literature to measure the burden of female reproductive conditions due to EDCs. They instead focused their work on effects of endocrine disruptors on other components of ODS, especially the notions that pesticides can contribute to fibroids and that phthalates can contribute to endometriosis.[306] Animal studies have shown that persistent organic pollutants, including DDT, can induce fibroids.[307,308,309] We know that DDT can induce effects on estrogen, which is known to induce growth of uterine muscle cells, the major cell type whose growth is dysregulated in fibroids. Yet the experts could not confirm this as the precise mechanism, and so they rated the laboratory evidence less highly than some of the other associations we've discussed.[310]

When the experts looked at the human studies on fibroids, at first glance the 11 studies looked inconsistent. The studies of fibroids and EDCs looked at different exposures. Some studies did not consider other potential risk factors such as race and age as alternative explanations, and their diagnostic criteria were different. When you hear of researchers pooling data across studies, sometimes they put apples together with oranges and get misleading interpretations.

The best of these studies was led by Germaine Buck Louis

and recruited women from 14 clinical centers in the United States, excluding women with endometriosis. This study looked at chemicals measured in the fat inside the abdominal cavity called the omentum, which protects and covers the intestines. As many organic pollutants are fat soluble, they accumulate there and persist for years if not decades. Given the proximity of the uterus to this fat, measuring concentrations of these chemicals here is arguably the best way to estimate past uterine exposure to these contaminants.[311] The other studies did not have fat samples that they were able to analyze.[312]

None of these studies had followed these women from their childhood or even earlier, when their mothers were pregnant. Future studies need to put these kinds of populations together to advance the science. Germaine's study did measure persistent chemicals, so the exposure measures likely can be interpreted to represent the relevant early-life exposures that precede disease. Fat samples were collected inside the abdomen of all the women who happened to be undergoing a laparoscopic exploratory surgery in the abdomen. The researchers compared concentrations of persistent organic pollutants in the omental fat of women with fibroids to those in women who had other diagnoses. A key breakdown product of DDT, DDE, was associated with a higher likelihood of fibroids. Reasoning from a single study has its limits, and so the panel rated the probability of a cause-and-effect relationship to be roughly a 1-in-3 chance.

The implications of those findings are substantial, suggesting that 37,000 new cases of fibroids requiring surgery will occur each year. For context, about 200,000 women undergo fibroid surgery each year in the United States. Just looking at the direct medical consequences, the cost would be $259 million.

FRACKING OUR HORMONES

Fracking is the process of obtaining oil and gas from deep inside rock formations by pumping sand and fluid to force open cracks in the rocks to access pools that have developed naturally under the earth. These techniques were advanced under the premise of energy independence in the United States, but as documentaries and newspaper articles have described, they have disrupted water supplies, reportedly making them flammable in some cases. With these techniques come large equipment and vehicles that bring with them noise and air pollution. But you probably didn't know that fracking uses more than 750 chemicals, many of which are EDCs.

Susan Nagel at the University of Missouri and Christopher Kassotis, now at Duke, have led the way in raising serious concerns about hormonal disruption and downstream health consequences, particularly infertility associated with fracking. They first went to Garfield County, Colorado, tested samples of surface water near fracking sites, and compared the hormonal activity to samples from control sites. They found higher estrogen activity (as well as activity against the female sex hormone) as well as antagonism to male sex hormones.[313] Then they exposed mice to these samples and found altered levels of pituitary hormones and reproductive organ and body weights as well as disrupted ovarian development.[314,315] They also reviewed 45 human and animal studies and found evidence for miscarriages, reduced semen quality, prostate cancer, birth defects, and preterm birth.[316]

Multiple studies from Pennsylvania and Colorado have identified concerns about prenatal exposure in families living near fracking sites, including birth defects and prematurity.[317,318,319] While it's difficult to say whether these consequences result directly from hormone effects, they raise serious concerns as fracking potentially expands further. And we haven't touched on the long-term impacts on wildlife and agriculture. While lower gas prices have slowed enthusiasm for new fracking sites, studies like these will help put the costs and benefits into better focus as policymakers debate whether to allow new sites to be fracked.

ENDOMETRIOSIS

The EDC Disease Burden Working Group benefited from increasing attention being given to chemicals and their potential contribution to endometriosis. The evidence against banned persistent organic pollutants such as PCBs was particularly strong across the 33 human studies we identified. We decided not to study these effects because no further regulation would help—after all, PCBs have been banned for 40 years in the United States. DDT has been banned, too, but we estimated the impact of DDT for fibroids because there are some places where it is still used for malaria prevention. There may be rare situations when DDT is the only alternative, but many studies suggest that safer alternatives can be used to prevent malaria, a deadly disease that is increasingly resistant to medication.[320]

There were seven studies that had examined phthalates and their potential effects on endometriosis. Not all of these seven studies met the high bar we set for inclusion in our analysis.[321] For phthalates and endometriosis, only one study met our threshold for rigorousness. Germaine's multicenter study, which I described earlier, had recruited 495 women undergoing laparoscopy, regardless of the indication for the surgery. Some women had endometriosis and others did not. In addition, the research team collected samples from a control group of women who were not undergoing such a procedure but were from the same residential area and of the same age. Consider this "control" group a second way to compare those who had endometriosis with those who did not.[322]

The study found consistent associations of phthalates both ways it looked at the data. The associations were stronger comparing the women with endometriosis with the control group. Exposure to a high-molecular-weight phthalate found in food packaging (DEHP) as well as another phthalate (dioctylphthalate) found in soft plastics was found to be higher among those with endometriosis. The study took a urine sample around the time of the procedure representing recent exposure, which is problematic when you consider that endometriosis can take months or longer to emerge. We need more studies that assess exposure at more than one time point, going back to when these women were undergoing puberty, or beyond, perhaps to when their mothers were pregnant with them. Accepting these as limitations, the panel was careful, deciding on a roughly one-third probability of cause and effect. In animal studies, DEHP has been found to affect ovarian development as well as embryo implantation, possibly through a mechanism that involves disruption of the endometrium, the uterine lining.[323] The findings in animals support the association found in the best human study available and the estimate of potential disease costs.[324]

Taking the best available US exposure data into account, we estimated 86,000 additional cases of endometriosis annually due to phthalates. When we calculated the cost of such an increase, we also took into account the lost quality of life that occurs in women with endometriosis over the 10-year period after diagnosis. Adding the direct medical expenditures to the other important indirect costs, we identified $41 billion in preventable costs due to phthalate exposures in women 20 to 39 years of age. Don't compare the costs of endometriosis to the costs of fibroids, as these are not apple-to-apple comparisons. These larger numbers speak to the reality that effects induced by EDCs on organs other than the brain can have huge social consequences that need to be considered in the public dialogue about prevention.[325]

Remember how I mentioned that we focused on one study when there were seven to begin with? Some have tried to simply add all the bad and good apple studies together to basically argue that there were no effects from the beginning.[326] In particular some of them relied on self-report of endometriosis, which is problematic because many endometriosis symptoms overlap with intestinal and other reproductive problems. Much as we would have liked to analyze data from all the studies together because it theoretically increases the power to detect a subtle difference, biased data can poison the mix and lead to misleading results.[327]

PREMENOPAUSAL BREAST CANCER

Breast cancer is a leading cause of mortality in women, behind lung cancer and heart disease. The Agency for Healthcare Research and Quality estimated an $80.2 billion in direct medical costs in the United States in 2015 attributable to breast cancer.[328] As difficult as the detective work is for fibroids and endometriosis, linking EDCs to breast cancer is arguably more difficult. There are many types of breast cancers. The biology of premenopausal and postmenopausal breast cancers vary both in their causes and their effects and prognoses. Some have estrogen receptors, meaning that they grow in response to estrogen and shrink when tamoxifen blocks its effects. Tamoxifen does not work as well on breast cancer cells that lack the estrogen receptor, which are called ER-negative.

The DES story shows clearly how EDC exposure can lead to breast cancer. We should also consider the story of hormone replacement therapy (HRT), typically with estrogen and progesterone used as medication to offset or treat the side effects of perimenopause and menopause. Though debate continues, for certain women, HRT can increase breast cancer risk within 5 years, particularly for women carrying genes that predispose for breast cancer.[329,330,331] Breast cancer is also associated with other risk factors, namely an early age at first menstrual cycle and never having given birth. What's the connection? The thinking is that more exposure to estrogen during menses surges may stimulate breast cells to divide and mutate their way toward uncontrolled growth, i.e., cancer. We do know that removal of the ovaries can dramatically reduce the frequency of breast cancer, including in women with BRCA mutations known to pose the highest risks.[332]

Barbara Cohn at the Public Health Institute in Oakland has done one of the most elegant studies of chemicals and breast cancer. Barbara and her team used serum samples taken from women just after giving birth and kept in contact with the women for nearly 30 years, following them up to see if they developed premenopausal breast cancer. They measured persistent organic chemicals, including DDT, in the serum samples. Because of the long half-life of these chemicals (ten years or more), the measurements taken just after the women gave birth are much better at estimating exposure during childhood than the previous studies, which had recruited women when they developed breast cancer. The researchers suspected that exposure to EDCs during breast development can fundamentally change the body in a way that sets up for increases in cancer risk. Barbara compared levels of DDT in the mothers who developed breast cancer to the levels of DDT in a matched comparison group of mothers without breast cancer. Women with the highest levels of DDT had a fivefold higher risk of breast cancer.[333]

A striking comparison also revealed more support for Barbara's findings. Among women who were under 14 years of age in 1945, when DDT use came into widespread use, the association with breast cancer was strongest. Among women who were born earlier and were not exposed before adolescence, there was no association. This finding speaks to the importance of breast tissue's susceptibility to DDT during childhood and puberty. DDT levels in women of childbearing age have declined substantially since the 1950s and 1960s (when the women participating in Barbara's study were recruited). That said, like Anderson Cooper, women of childbearing age today frequently have detectable levels of DDT. And don't forget that DDT is still used in some parts of the world to combat malaria. Barbara and her colleagues have developed sophisticated models that allow us to estimate exposure in women now so that we can determine the implications of exposure that potentially persist to this day. Using Barbara's data, a "back-of-the-envelope" calculation suggests that as of 2010, as many as 14,900 cases of breast cancer in Europe could be due to DDT exposure, with costs that could be as high as €685 million. The numbers in the US are likely similar.

ATRAZINE AND LETROZOLE – A SAD IRONY

Atrazine is the second most widely used herbicide in the United States, behind glyphosate (Roundup).[334] It's primarily used to control weeds in corn crops across the Midwest. Atrazine turns on aromatase, an enzyme that transforms testosterone into estrogen. Interestingly, the same company that makes atrazine also makes a chemical called letrozole that inhibits aromatase and is used in some breast cancer treatments. Certain breast cancer cells carry an estrogen receptor and grow in response to estrogen when the estrogen binds to this receptor. Whether you call letrozole treatment or an endocrine disruptor by design, there is substantial irony here.[335]

POSTMENOPAUSAL BREAST CANCER

The estimate of 14,900 breast cancer cases in Europe due to exposure to DDT, which we have not published, is limited to premenopausal breast cancer, which onsets before age 50. Another study, in Spain, focused on postmenopausal breast cancer, collecting fat samples from breast tissue in women undergoing surgery for cancer, as well as fat samples from a group of comparison patients of the same age and from the same hospital who underwent surgery for other reasons. Many of the studies we've talked about previously in this book have focused on individual chemicals, but the reality is that we live among thousands of chemicals, each with different influences on estrogen and other hormones. And so, rather than test samples for individual chemicals, the researchers measured the samples for their estrogen activity and compared the results. Among women with higher levels of estrogen activity, breast cancer risk was higher, concentrated among women with lower body mass index. A potential explanation for the difference in effect by BMI could be that adipose tissue can soak up these chemicals, which also tend to be highly fat-soluble, and protect the breast from effects of environmental estrogens.[336]

A follow-up study from this group in 2016 looked at a larger and different population of Spanish women and was more representative in its approach, reducing concerns about bias and enhancing the generalizability of results to the Spanish population. That study used serum samples from women with breast cancer and another matched comparison group. In this study, environmental estrogens (as assessed by measuring the estrogen activity of the serum) were also associated with increased risk of breast cancer.[337] Extrapolate the data from the first of these two studies of postmenopausal breast cancer, and in Europe you get 70,600 additional cases each year with costs of €3.25 billion attributable to chemical exposure.

The limitations to these as-yet-unpublished data will sound familiar. These are single studies measuring chemicals mostly at the time of diagnosis (except for Barbara's elegant study), when earlier in life is when the hit-and-run is most likely to have happened. Not all the pieces of the puzzle were put together to definitively say yes or no. The chemicals studied here stay in the body for years if not decades, making the study easier to interpret despite the limitations. We need more research in this area so that we can move beyond looking at older chemicals and focus on newer chemicals that may also be problematic. We're going to need to find markers of breast cancer risks and project based on those rather than breast cancer incidence. Otherwise we'll be 40 years behind forever.

Where does that leave our young patient, Emily? All hope is not lost. There are many other risk factors that coalesce to make girls and women suffer from early puberty, endometriosis, fibroids, and breast cancer. We know that exposures across the life course matter, not just in pregnancy, so that if Emily reduces her exposure going forward, she may be able to reduce her risk for these diseases. I do hope you've also taken away from this chapter the commonality that these conditions share: the disruptive, multifocal force that synthetic chemicals can apply with different consequences in each individual. I'm especially concerned about the broad impact these chemicals have in developing countries, which by 2030 will produce and continue to use the majority of synthetic chemicals.[338]

With that said, we can start now to transform our lives one by one, or hundreds by hundreds. And that includes raising your voice to be heard. I shifted my focus to public health and policy because I saw that I could do more to effect change on a population level than with my prescription pad. So let's change our focus from the bad news that these studies bring to positive change in our homes, schools, workplaces, and society at large.

PART THREE

TAKING ACTION

CHAPTER SEVEN

REAL STEPS THAT MAKE A DIFFERENCE

You might be worrying that the only way to protect yourself from the increasing prevalence of chemicals in our world would be to move to a remote farmhouse somewhere in the countryside, thousands of miles from any city or suburb. You might even be wondering how I can do work in this field and not get depressed. My reality is quite the opposite: I'm an optimist. I know that we can do a lot to improve our lives now and that we have the power to change manufacturing and industrial agriculture for the better. Below, I've gathered many of the "what you can do now" suggestions that I've hinted at throughout the book. These steps can make a real difference not only in protecting yourself and your loved ones from the downstream effects of chemicals but also in fundamentally changing the way companies make the products we use everyday in the first place. You can take safe, simple steps to limit exposure to the endocrine-disrupting chemicals of greatest concern.

LIMIT YOUR EXPOSURE TO PESTICIDES

While environmental regulation in the United States tends to lag behind Europe, pesticides are an example of our doing the right thing sometimes. In large part due to the Food Quality

Protection Act, which requires an additional safety factor to protect children from pesticides in food, organophosphate residues in American food have come down somewhat. Indeed, levels of organophosphate breakdown products in the urine of children and pregnant women are lower in the United States than in Europe.

Bans in organophosphate use in households have also benefited children in the United States, as we saw with the study at Columbia University that showed lower levels of the pesticide and no relationship to birth weight and length after the chlorpyrifos ban. This is good news.[339]

So we've made some progress, but aside from banning pesticides, what can we do to prevent our exposure to them? Eating organic has been documented to reduce levels of organophosphate breakdown products in the urine.[340,341] There are certain vegetables and fruits that are known to have the highest levels of pesticides that can get into people's bodies (see box on page 137). For some of these fruits and vegetables, taking off the rind is an option, but for others, even washing the fruit carefully gets off only so much of the pesticide residue. Other vegetables have been found to have low pesticide residues, including asparagus and cauliflower. The Environmental Working Group has a great website (ewg.org) and app if you're interested in learning more.[342]

Many people are concerned about the costs of eating organic. I work at Bellevue Hospital, the flagship of the public hospital system for New York City, where I encounter many families who are working extremely hard just to make ends meet. As I described in chapter 3, changing to an organic diet can reduce organophosphate metabolites in the urine of children, even those in low-income urban and agricultural communities.[343] More and more stores such as Costco, Sam's Club, Walmart, and Target are offering organic produce and foods at reasonable prices. You can grow your own vegetables and

THE TERRIBLE 12

According to the EWG, these 12 fruits and veggies are especially vulnerable to absorbing chemicals and therefore pose a greater risk when you buy "conventional" rather than organic produce:

Strawberries	Spinach	Nectarines
Apples	Grapes	Peaches
Cherries	Pears	Tomatoes
Celery	Potatoes	Peppers

fruits using organic gardening practices, or join one of a growing number of community supported agriculture (CSA) programs across the US. Farmers' markets are increasingly common. In New York City, Supplemental Nutrition Assistance Program money (also known as food stamps) can be used in its farmers' markets—and can get bonus money for buying there!

Eating organic also has other benefits. The organic label means the food does not contain genetically modified organisms (GMOs). While debate continues about the health effects associated with genetic engineering, some pesticides used with GMOs have been shown to disrupt hormones. The science here is still emerging but worth keeping an eye on.

In 2016, President Obama signed into law legislation requiring the US Department of Agriculture to set rules for disclosure regarding GMOs. The USDA released its guidelines in May 2018, and they confuse the public by insisting on using the term "bioengineered," with a logo of a brightly shining sun.[344] Done correctly, GMO labeling would be a solid step toward giving people the option to pick and choose for themselves.

LIMIT EXPOSURE TO PHTHALATES

It may seem obvious, but one easy way to reduce exposure to phthalates is to eat fresh foods. One study that drove this point home followed five families as they changed from a regular diet to a "special" one free of canned foods and prepared almost exclusively without contact with plastic. Levels of phthalate metabolites, specifically DEHP, dropped between 53% and 56%. When participants changed their diets back to normal, levels came right back up.[345]

Using glass containers avoids any concerns altogether, but sometimes that's not realistic. Many schools do not allow glass containers for security and safety reasons. Consider using stainless steel. If you must use plastic containers, use them correctly:

- If plastic containers were meant for single use, don't reuse them. In addition to EDC risks, reusing them raises the chance of bacterial contamination.
- Look at the recycling number on the bottom of the plastic container. The number 3 means phthalates, which raises the possibility of contamination into liquid or food.
- Never microwave plastic. You're inviting plastics to melt at a microscopic level and travel into food. There is no such thing as microwave-safe plastic.
- Never wash plastic in the dishwasher. Handwash with mild soap and water instead. Harsh detergents etch the plastic and increase absorption into liquids and foods.
- If plastic food containers are etched, it's time to throw them away. Etching increases the odds of leaching.

You'll notice I haven't mentioned cosmetics and fragrances much, even though many of these products contain phthalates. That's not to say that there aren't concerns that lower-molecular-weight phthalates found in these products disrupt sex steroids in our bodies. The good news is that a number of companies have pledged to remove phthalates from their lotions and creams, working closely with organizations such as the Campaign for Safe Cosmetics. The Environmental Working Group maintains a comprehensive, highly accessible database called *Skin Deep* (ewg.org/skindeep) that distills what we know about ingredients and recommends cosmetics that present the fewest risks.[346] You can also look at the ingredient label and avoid products with "fragrance" or phthalates. Nail polish, hair sprays, and deodorants frequently contain phthalate ingredients. A recent study of young girls found that choosing personal care products that are labelled to be free of phthalates, parabens, triclosan, and benzophenones can reduce personal exposure to possible endocrine-disrupting chemicals by 27% to 44%.[347]

LIMIT EXPOSURE TO BPA

There are two main avenues for BPA exposure in humans: canned food and beverages and thermal paper receipts. Of the two, diet is the most problematic route of exposure, especially in children, for which it represents 99% of exposure.[348] Further support for this notion comes from that study I described in which families changed their diet to fresh foods, which also saw a 66% drop in BPA levels.[349] Another study did the opposite, asking people to eat canned soup multiple times each day: levels shot up over 1,200%.[350] Stopping canned food consumption can drastically decrease BPA levels in urine, as much as 90% or more. While certain levels of acidity can make BPA leach less or more, BPA gets into all foods pretty well, whether it's canned soda or vegetables.[351]

You do have some options and safer alternatives to BPA-containing cans. Food packaged in Tetra Paks avoids the use of cans altogether while avoiding food-borne microbiological disease. Some companies are increasingly using a naturally derived lining called oleoresin, which is modestly more expensive (2.2 cents a can) compared to the polycarbonate resin currently used. As I mentioned earlier in the book, we tallied up the potential costs and benefits of replacing BPA with an alternative free of health effects, and found nearly an equal tradeoff. In some scenarios that considered the uncertainties in our calculations, the benefits of replacing BPA were greater than the costs.[352]

Nevertheless, there remains a lot of resistance to banning BPA in canned foods and beverages. BPA is increasingly being replaced with a structurally similar chemical called BPS, substituting a sulfur (S) where a key carbon atom is placed in BPA. BPS is more persistent in the environment and appears to have similar estrogenic potency.[353,354,355,356,357,358,359] BPS is also replacing BPA in thermal paper receipts, the other route of exposure. Until recently, BPA was also being used in plastic water bottles; you could tell which bottles had BPA because they had the number 7 inside the recycling symbol on the bottom. An extra step to reduce exposure to BPA and BPS is to buy bottles that say BPA- and BPS-free on the label or to avoid bottles with the number 7 altogether.

LIMIT EXPOSURE TO FLAME RETARDANTS

As I discussed in chapter 3, the science on flame retardants should not make you run to throw out all the furniture in your home. The good news is that you can reduce your exposure to these chemicals in your home right now through some simple steps:

- Replace old furniture that has exposed foam, or cover it with a slipcover.
- Buy products made from natural fibers (like wool), which are naturally less flammable.
- Open your windows! Outdoor air has lower concentrations of flame-retardant chemicals, and recirculating the air a few minutes every day gets rid of other chemical residues, too.
- Vacuum regularly with a HEPA filter and mop with a wet mop to prevent contaminated dust from indoor electronics, carpeting, and furniture inside and outside the home from accumulating.
- Stop children from touching fire-retardant items or putting them in their mouths.
- Make sure you get a healthy diet with enough iodine. In 2007, the World Health Organization reported 2 billion people worldwide have insufficient iodine intake.[360] Iodine is critical for thyroid function.[361] Seaweed is one of the best sources. Seafood and dairy products are other good sources, as are cranberries and strawberries.

Now, I must provide some caution about the evidence behind these suggested steps. Interventions do not always achieve the expected results. These are not failures but speak to a broader problem we will address in chapter 8. Sheela Sathyanarayana, the pediatrician-researcher from Seattle who has been mentioned multiple times in this book, conducted an intervention in families to reduce phthalate exposures by changing diet that produced unusual results. Phthalate levels were much higher in the intervention group, baffling everyone! After painstaking work to comb through every ingredient in the participants' diets, she found that the coriander given to participants had been ground in a way that plastic particles were inadvertently contaminating this spice, producing the spike in phthalate levels that you would have expected from eating highly packaged foods.[362]

We assume that we can duplicate the laboratory in real life and control every source of exposure. While we can control many, findings like these speak to the need for systemic change.

Another example of a misunderstood "negative" finding is a citizen science study conducted by Tamara Galloway and colleagues at the University of Exeter. They worked with students to develop a scoring scale so that people can evaluate BPA contamination and limit exposure themselves. Exposures were low to begin, and the researchers didn't measure bisphenol replacements like BPS and BPF.[363] Urine levels of BPA were not significantly different during the intervention compared to before the intervention. Bad design? No. Bisphenols are pervasive not just in foods but in thermal paper receipts, polycarbonate plastics, and dental sealants.

Focusing on the home may be too myopic an approach. Just ask one of the leaders in exposure science and measurement of endocrine disruptors, Kurunthachalam Kannan of the New York State Department of Health. He's probably one of the humblest people I know in this field, yet his presentations describe a scary reality. "We are eating a half a milligram of plastic each day," he explains. "You won't see it the way you see fish with plastic bottle caps in their intestines, but the same plastic is there, just as microscopic or smaller particles." He presents data on the levels of phthalates he has measured in various foods purchased in supermarkets, personal care products, and household dust. He estimates that these sources explain only one-fifth of the total internal dose based upon the measured levels in people. So where are all the hidden phthalates? Think about the workplace, the cars or subways or buses we use to go to work or school, or all the other places we visit in our daily lives. We'll talk more about limiting exposures in all these other environments where we experience EDC exposures in the next chapter, but suffice it to say that it's remarkable that studies have reduced exposure just through a household intervention! Imagine what would happen if we took on a broader approach to eliminating these exposures in the first place.

Should the ubiquity of chemicals in our midst and the controversial, complicated nature of studying their dangers deter us? No! It's remarkable under these circumstances how many interventions have actually worked and how well they do. The success of interventions speaks to the need for broader change. Many of these steps focus on the home, but many of us spend only weekends and evenings there. Our workplaces, schools, subways, buses, cars, restaurants, and other public spaces all can influence our exposure. I do not believe we can rely on policy alone to fix these problems. Let's talk about the broader movement that's necessary. We need to use our economic power to change the system for all of us to benefit.

MEDICINE IN 2040

Consider for a moment how things might change if awareness caught up with the reality of endocrine disruption. Let's imagine that Diana and Eduardo are planning a pregnancy. They go to their primary care providers for checkups. They have unremarkable medical histories and have had few previous risk factors that would put their fertility at risk. They are both middle children, and their parents were healthy and had them young by today's standards, in their mid-20s. Diana and Eduardo are both in their mid-30s and want to have as much data as possible to inform their decision. With what we know so far, they would have a 91% chance of conceiving in the first year.

In pursuit of a more personalized assessment, Eduardo goes to Dr. Sanchez, a urologist, for a fuller evaluation. Dr. Sanchez orders a semen analysis, which reveals Eduardo has a normal count of motile sperm. The bad news is that he scores in the 90th percentile for exposure to BPS and DINP. (DEHP was banned in 2025.) His serum tests suggest minimal exposure to flame retardants and the other persistent organic pollutants.

His urologist asks whether he has been eating any processed foods. "Actually, I've been on a diet that focuses on greens, beans, and protein. I've lost twenty pounds!" Eduardo replies. Further questioning uncovers that Eduardo has a stash of canned beans in his office above the microwave. Dr. Sanchez suggests he get retested after switching to Tetra Paks or glass containers. "And don't forget to eat organic!" he advises. A few weeks later, Eduardo's BPS level plummets, and he's good to go from Dr. Sanchez's perspective.

Diana goes to Dr. Smith, a reproductive endocrinologist, who orders estrogen and progesterone measurements and performs an ultrasound. She also runs an epigenome-wide sequence in addition to a genome sequence. This way, she can look at gene expression as well as gene sequence and see two types of risk factors for disease. We've come a long way since 23andMe! The report suggests overexpression of some genes and hints at elevated BPS levels, and her urinalysis confirms just that. While her hormone levels and ultrasound are fine, Dr. Smith suggests a focus on folate-rich foods to push the genome expression the other way.

We're a few technological leaps and bounds from this world. It's expensive enough to sequence your genome; gene expression testing for your whole genome is cost-prohibitive, and we don't know yet how to apply epigenomics in clinical care. The Precision Medicine Initiative is a large-scale study of a million Americans that could jump-start the effort to identify early predictors of disease and fundamentally transform clinical care. It might even be able to use a pregnancy test kit capable of measuring EDCs with some help from a mobile device.[364] We need technology to catch up. Of course, there are other "-omics" that may progress to being used by patients, empowering them to guide their own health.

You've noticed I've focused on a couple thinking about having a family. All of the pregnancy studies I describe may not have measured at the right time, when exposures can impact children's health. Keep in mind, we are born with the germ cells that eventually become sperm and ova. By the time the pregnancy test changes color, the embryo that forms from sperm and ova has already been dividing. Soon after fertilization has occurred, the heart begins to beat. So in some respects, most pregnancy studies have missed the most relevant exposures. There are a number of research groups examining studies of couples intending to conceive and then following their children. Stay tuned for even more information as these studies launch.

The other interpretation of the basic biology is that all exposures at every life stage matter. People should be thinking about prevention even if they are not planning a pregnancy. Remember the effects on the male reproductive system discussed in chapter 5? We don't have specific evidence to prove that sexual performance is inhibited by EDCs. However, given what we know about EDCs' effects on sex steroids and the role of sex steroids in sex drive in men and women, there might be immediate benefits from avoiding exposure. Perhaps you can be bold enough to ask about the materials used in your workplace, as we'll talk about in the next chapter.

CHAPTER EIGHT

YOUR VOICE MATTERS: HOW YOU CAN PARTICIPATE IN A VIRTUOUS CIRCLE

I teach undergraduates about environmental health at NYU. The way I structure the course is highly unusual, because I spend an entire lecture every semester discussing economics. I begin my lecture by talking about Adam Smith, the father of modern economics. Smith described the features of a properly working market economy as being led by an "invisible hand" that directs the best possible market because it assumes all buyers and sellers are self-interested. Smith believed that government intervention could induce market dysfunction, ultimately creating more harm than good. In today's day and age, then, he might be considered a libertarian, against government regulation.

So what would Adam Smith say about endocrine-disrupting chemicals?

I believe that he is probably spinning in his grave right now over this issue! He would be upset that the manufacturing of these chemicals and their ubiquity in the products we use both create market externalities. (I promise, I am not trying to write *Environmental Health Freakonomics*, and I'm not a PhD-trained economist, but bear with me for a few minutes.)

THE ECONOMICS OF ENDOCRINE-DISRUPTING CHEMICALS

Adam Smith's principle was that, when markets work well, only buyers and sellers can benefit (or be harmed) in a transaction. Secondhand cigarette smoke is a classic example of what economists call an externality. Cigarette smoke causes health and economic damage (the externality) to third parties who neither buy nor sell cigarettes. This damage is not properly accounted for in the price, so more cigarettes are produced than should be produced and get sold for less than society actually pays. The same could be said for coal-fired power plants. Coal is produced, sold, and used without taking into account the effects on children's lungs, including asthma exacerbations requiring hospital trips. In the case of EDCs, pesticide spraying in an agricultural community might cause pregnant women in neighboring houses to be exposed, damaging the brains of children and their mothers who are not working on the farm or otherwise financially benefiting from the growing of crops.

Similarly, the effects of EDCs in products are not considered when contaminated items are bought or sold, and the price of these products does not properly account for the degree of contamination—a result that can occur when the seller knows about pesticide dangers, for example, and doesn't disclose. Such a scenario is an example of asymmetric information.

Now let's return to Adam Smith and his theory of a world of perfect information, with buyers and sellers knowing everything about what is bought and sold. He was a big fan of transparency. When these economic principles fail, we don't have optimal economic productivity as a society. Governments can fix these kinds of problems, but they often trip over themselves when intervening. Sometimes government intervention works beautifully—at the beginning of this book, we talked about the phaseout of lead in gasoline as just such an example. The "tax refund" of $1,000 per person that we still receive to this day is actually the removal of a $1,000-per-person externality (or tax) placed by the lead and gasoline industries on children exposed to chemicals that damaged their brains year in and year out. Phasing out toxic exposures like lead is a victory for children and the economy.

British economist Arthur Pigou described taxes as a solution to market externalities. In theory, taxing industry could work, but for chemical exposures, the devil is in the details. You have to know the exact amount of damage the exposures are inflicting. For Pigou's idea to work, people suffering from diseases resulting from exposure receive the exact amount of damages from the taxes directly. That didn't even happen with the tobacco settlement money, as many states simply funneled that money into the general treasury. So you can understand my doubt that industry would agree to these kinds of paybacks or that they would work.

The bottom line is that the free market does not account for the effects of EDCs on people's hormones and ultimately their health. This costs money—hundreds of billions of dollars in the United States and Europe, if you add up all the costs identified by the expert panels we assembled back in 2014. When you total the costs of diseases due to EDCs described in this book, you find a total of $400 billion each year.

Simple addition is misleading, however, because the evidence for each of the diseases varies so widely. Discount the costs in relationship to the evidence and probability of causation, and you get a total likely to be $340 billion each year in the United States and $217 billion in Europe.[365] That number would give Adam Smith heartburn. Keep in mind that we examined fewer than 5% of all known endocrine disruptors, a subset of diseases associated with each EDC, and a subset of costs associated with diseases we connect with EDCs. The problems and their costs are much, much larger.[366]

LEVERAGING YOUR ECONOMIC POWER

Would Adam Smith go calling his congressman or state senator? Probably not. (He was Scottish, after all.) But you can.

The current federal climate is not very promising for progress on this front, and it's not helped by former EPA administrator Scott Pruitt's decision to consider only those studies that can make all their data available publicly in the policy-making process. For researchers like me, who follow mothers and children from pregnancy to childhood and beyond, federal privacy rules and ethics make this impossible and inappropriate. Pruitt couched this new rule as part of a transparency initiative, but it's an industry trick play. The Pruitt rule would leave only very small and largely industry-funded studies on the table.[367]

As I mentioned earlier in the book, when it comes to chemical hazards, states have led the way in pushing the regulatory agenda forward; however, this progress is slow and spotty. Policy can work, but for us to make continued momentum and real change, we are going to need to speak up for ourselves. You are the consumer, and you have the power to drive manufacturing with your wallet. Indeed, the power of the consumer is the secret behind my optimism in my work.

The market for organic food is a great example of this phenomenon. The science suggesting effects of pesticides on health was arguably weaker in 2010 than it is now, yet between 1996 and 2010, sales of organic foods rose in the United States from $3.5 billion to $28.6 billion.[368] As demand increases for organic food, economies of scale have the potential to reduce the costs per unit production and, in turn, the costs of organic food. This kind of "virtuous circle" has produced reductions in the levels of pesticides measured in people's urine. Reduced exposure in pregnancy means fewer IQ points lost and greater economic productivity in each cohort of children born.

Ten years ago, phthalates weren't on the radar as prominent consumer issues. As studies emerged suggesting possible effects on puberty, brain development, obesity, and diabetes, people started demanding phthalate-free water bottles. Data from national surveys conducted over the past decade by the Centers for Disease Control and Prevention found some remarkable drop-offs (17% to 42%) in urinary phthalates beginning in 2001 and continuing through 2010.[369] Over the same time span, burdens (and costs) of disease like those we described for infertility, obesity, and other conditions due to phthalates may have dropped.

Another example is removing BPA from baby bottles. As media attention was placed on studies in humans that confirmed the concerns first identified in the laboratory, consumers began to ask whether BPA was in the water bottles they were using at the gym or on the go. Between 2003 and 2010, measured levels of BPA dropped by 50%.[370] BPA was banned in baby bottles and sippy cups in 2012.

A QUEST FOR REAL TRANSPARENCY, OPTIMISM, AND ACTION

We talked a lot in the last chapter about intervention studies that have documented the possibility of lowering exposures to EDCs. These studies did the right thing by intervening in homes, where we spend most of our time away from work and school. Our homes are the first place we think of when we want to take steps to ensure our health and safety. And while our autonomy over the home environment has limits, it's substantial. We can select our food, prepare it, and serve it on our dining room tables. We can purchase personal care products and cleaning materials. But exposure to EDCs occurs in many more places where we spend large parts of our lives: at work, in schools, as we travel through airports and train terminals. And the data from Kurunthachalam Kannan and his colleagues mentioned previously prove the pervasiveness of exposures and supports this notion.

Most adults, for example, typically spend 8 or more hours per day at work, at least 5 days per week. Just picture your daily lunch scene. Some of you may have the time to prepare your meals at home and bring them to work; if you do, count yourself lucky enough to eat clean, safe food. But I would bet that most of us, me included, barely have time to leave our desks or have schedules so jam-packed with meetings that we can almost forget to eat. If there's a cafeteria, then there can be meals that resemble home but perhaps not with the same components, especially if you shop for organic fruits and vegetables.

Fast food is a common solution for the problem of the busy workday, but you should know that studies have found that eating fast food is associated with much higher (20% to 40% higher) levels of phthalates in the urine.[371] The source of these phthalates is likely the food packaging, which may more readily leach these chemicals due to the high temperatures at which the food comes into contact with plastic materials.[372] Given what we already know about fast food and its calorie content, plus what you've learned about phthalates and their disruption of metabolism, there may be more than one reason not to pull over and grab that greasy burger to go.

Now I'd like you to think about the physical conditions of your workplace or your child's school—its furniture, its lighting, its paint, the soap in the bathrooms, the cleansers being used to clean after hours. It might seem we have little control over the places we work. While we do not decide what materials are used, our employers have even more control than any one of us does when we are shopping for our homes. It doesn't make you troublesome colleagues to organize a "green team" to bring up the strong smell of the air freshener used by the cleaning staff at night. You might ask building managers to minimize the spraying of pesticides on the grounds around the building, inside the building, or on the school playground.

In California, teachers in the Solano Unified School District partnered with the district's administration and insisted on safer materials and cleaning practices. It ended up saving money, too! Palm Beach County's school district in Florida estimated $360,000 in annual savings by implementing a green cleaning program in 180 schools in 2008. And those estimates don't fully consider the health benefits that come from reductions in asthma and other illnesses among students and staff. In California, that's $40 million each year.[373]

Even if you don't quite believe in individual consumer power, think about corporate power! If a large company's employees all spoke up, that company would have a lot of leverage to force a major supplier to find out and show how and with what ingredients their products are made. The National Basketball Association, for example, has partnered with the National Resources Defense Council to create a Greening Advisor website for its 30 teams. Its website includes information about integrated pest management programs to reduce pesticide use and limit use to the least toxic pesticides. There are also sample letters for teams to send to suppliers in an effort to encourage them to do the right thing.[374] While I wouldn't say the website is perfect or covers all the EDCs we described in this book, this example speaks to momentum and potential. After all, Forbes now values each NBA team at over $1 billion.[375]

Whether it's at home or work, my message here is this: you have more power to manage your environment than you think.

LABELS AS A MECHANISM FOR INNOVATION AND POSITIVE DISRUPTION

Laws won't be changed quickly enough to protect people, and you shouldn't have to wait so long. Remember how BPA was banned from baby bottles and sippy cups? By the time the FDA banned it, most if not all manufacturers had already changed their practices so that they could place "BPA-free" on the label.

Of course, we can advocate for policy change, and when states and countries do the right thing, we should hold companies accountable. When California finally rolled back its requirement to apply flame retardants to furniture, companies didn't simply take them out. Some prominent advocacy organizations helped this process along by putting this issue on the media's radar. And guess what? Furniture companies saw the opportunity to protect their market share and proudly documented the absence of these chemicals in their products.[376]

In an ideal world, you should also ask for data to confirm the absence of contamination. I tell the undergraduates I teach about my hope that someday the nutrition labels on foods will list not just the protein, carb, and fat contents but also information about contaminants. Some makers of omega-3 fatty acid supplements already document the absence of mercury and persistent organic pollutants (POPs) in their products. Some will argue that my suggestion is ill-fated, arguing that the public won't understand or will misinterpret a detectable level of an organophosphate pesticide in foods. I appreciate the concern, but the need to show the data can be highly motivating for manufacturers and provide a new opportunity for healthy competition and innovation.

Will this require costly testing for chemicals in batches of food? At first, perhaps, but the costs of environmental protections always seem high and even inflated when they are first proposed. As demand rises enough to develop processes that would bring the testing onto a larger and cheaper scale, prices will drop.

In graduate school, I had the opportunity to take a class on innovation at Harvard Business School that brought this concept into focus for me. Originally, having read the work of Clayton Christensen, I was focused on innovation in health care delivery. Christensen coined the phrase *disruptive innovation* back in 1995 to describe innovation that fundamentally changes the value system, creating a new set of winners and losers. Think about the shifts from floppy disk drives to CDs to flash drives, or Amazon with retail sales for everything under the sun. Companies with established market share can have their world turned upside down despite major investment in research and development and good customer service. Upstart companies that have developed a new technology end up winning market share by designing a product that resets the basis for competition, sending established players into the red.[377]

These events disrupt the market. Established manufacturers may lose market share because they have failed to adapt to change, sticking to using the ingredients and processes that drove them to the top to begin with. They may resist replacing ingredients that may expose people to EDCs because they have to rework their supply chain. They will complain that it's costly—but they will simply pass that cost to the consumer. Consumers win when disruptive innovations for all sorts of products reset the competitive landscape for product characteristics, price, and other factors. Some established manufacturers may lobby for protection from government to protect against disruptive innovation. Adam Smith would find this idea repulsive because we all stand to gain from disruptive innovation. When these disruptions happen, the market becomes more efficient, providing greater value to the consumer.

Are we seeing a series of disruptive innovations in the market for food? Substantial investment has yielded a number of efforts to fundamentally change the way we produce, consume, and transport food. The rise of organic food is just one driver of this phenomenon and is itself a force multiplier toward innovation. Organic food sales continue to show near-double-digit annual percent gains, whereas the overall food market was stagnant in 2016, growing less than 1%.[378] While organic is still only 5% of the food market, as that percentage increases, the market share will drive technological and other innovation, reducing cost and enhancing affordability. You can see the impact when big-box stores such as Costco and Walmart, and Amazon through its acquisition of Whole Foods, have started to offer organic options. Organic food can become a regular part of the 99-percenter lifestyle and is no longer simply for the well-to-do.

Beware marketing masquerading as healthy food. Take "natural" labeling for example. This is one of the most unfortunate loopholes in food labeling. "Natural" says nothing about pesticide or phthalate contamination. The Food and Agriculture Organization, the United Nations body overseeing food and agricultural production worldwide, collects international standards for food. Though one exists for organic food, no such designation guides what is so-called natural food.[379,380] The same goes for the US Food and Drug Administration. For a "natural" label, the FDA requires only minimal processing and the absence of added ingredients or color.[381] Though the Federal Food, Drug, and Cosmetic Act prohibits misleading labeling, the absence of a "natural" definition is confusing at best and at worse misleading.

The same goes for the "free-range" designation when applied to meat and poultry. In terms of the dangers from synthetic chemical contamination, poultry and eggs labeled "free-range" are not necessarily free from pesticides. The USDA simply requires that "free-range" eggs are "produced by hens housed in a building, room or area that allows for . . . continuous access to the outdoors during their laying cycle."[382,383] Buying free-range may make you feel good if you hate factory farming, but that's a separate issue.

THE LACK OF TRANSPARENCY

There is another major obstacle to changing our exposures to EDCs. Many manufacturers use patents and "trade secrets" arguments to avoid disclosing the ingredients in their products. On this front, California has made some legal progress. The state now requires disclosure of ingredients found in cleaning products. In many cases, no new chemical hazards may be hiding in the special flavors or scents or other additives in the carefully concocted mixtures of known chemicals, but without transparency, our health and economy remain at risk and we don't even know it.

Foods are another area where chemical additives can be used without full and proper toxicity tests. Under the Federal Food, Drug, and Cosmetic Act, an industry scientist can vouch for the safety of these additives under a Generally Recognized as Safe (GRAS) designation. The FDA can't require proof of particular safety tests when GRAS is invoked. Of the 3,941 food additives listed on the FDA website "Everything Added to Food in the United States,"[384] reproductive toxicology data were available for only 263 (6.7%), and developmental toxicology data were available for only 2. I think most people would be disturbed, if not outraged, to discover that a chemical or product is considered "safe" because one scientist working at a company said it's safe.

I am certainly not suggesting that the majority of additives approved for use in foods is unsafe. And I am not suggesting you need to learn about each of the ingredients on the label. You will do plenty of good by limiting the exposures we've talked about throughout this book and focusing there.

What we can do, however, is to insist on transparency about what is in food or, for that matter, any product we purchase. That creates an opportunity for experts to evaluate what's in these products and ultimately identify and fill the knowledge gaps. The information doesn't have to create a huge burden for companies; websites, quick response (QR) codes, and other ways of presenting information can simplify product labels to avoid overwhelming people.

Some will say that, in this globalized market, companies can't get their partners in other countries to document correctly what ingredients are in their products. They might argue that they can't get cheap manufacturers abroad to change their practices for the better. Could the kind of competition I describe foster a greater American presence in the marketplace? Insofar as producers who are more careful about contaminants are based in the United States, wouldn't that bring jobs back to America? We've seen how manufacturing and policy practices can drive exposures, contribute to disease, and cost our country in the form of economic productivity. If we could reverse this problem and bring jobs back to our country in the process, that would seem to be a change welcomed across the political spectrum.

EDCs are a global problem, too, as are the health problems we've identified in this book. If we don't do the right thing, companies in other countries might recognize the threat ahead first. The countries that lead the way in making the changes will eventually win by being more competitive in the eyes of the public that increasingly values avoiding EDC exposures.

WHAT MIGHT WE EXPECT IN THE FUTURE?

Could GM feed its employees only organic food? Will Kaiser Permanente use phthalate-free medical equipment? Is there a time when Whole Foods purchases packaging only from suppliers that document absence of PFCs, perchlorate, and flame retardants? Might Amazon even go so far to offer discounts on foods in glass containers and insist on the use of phthalate- and bisphenol-free food packaging?

That future may not be so far off. Ostensibly to improve environmental stewardship, the Kraft Heinz Company announced in late July 2018 that 100% of its packaging will be recyclable, reusable, or compostable by 2025. While nowhere in this commitment is an explicit statement about chemical contamination, it wasn't that long ago that a movement began to push large multinational corporations towards a "circular economy" in which waste is minimized by closing and narrowing production processes and maximizing reuse. We'll see if Kraft Heinz actually does what it promised. If it can change the way it packages food, then the impact could be enormous.[385]

CHANGING THE HEALTH CARE MINDSET

It may seem as though the changes suggested in this book will produce benefits only in the long term, but consider the short-term benefits that come with reducing exposure. If you stop using fragranced products that contain phthalates, you will potentially see changes in the levels found in your urine within 24 to 48 hours. Phthalates are known to irritate the respiratory tract, so if you have allergies or other sinus difficulties, you may feel better even sooner. You could see changes in sex hormone levels, which have a typical half-life in the human body of about a week, not that long afterward. The effects on obesity and diabetes no doubt will take longer, but the other potential benefits will make it worthwhile even sooner.

So why aren't health care providers talking about this issue at every medical checkup?

First and foremost, there's no such thing as a prescription for prevention. Medical doctors are trained to diagnose and treat. Prevention doesn't get the same—or deserved—attention in medical education. Focusing on environmental health, the gaps are even more obvious. Many medical schools offer no or minimal training in environmental health. Back when I was in medical school, the average student received 7 hours of environmental health training in their entire 4 years of school.[386] That's less than the time it takes to do a typical coronary artery bypass operation—an experience common in surgical rotations in the third year. Residencies are not much better—a 2003 study found that fewer than half of pediatric residencies covered any topics other than lead exposure and environmental factors contributing to asthma.[387]

When it comes to environmental health, an overhaul of the medical school curriculum is long overdue. No surprise then that some doctors might shy away from talking about this—after all, three of the most dreaded words in health care are "I don't know." But if we insist on raising these issues and asking, it will force even the more senior doctors to read up and be more prepared to address them. That demand will also provide a stronger stimulus for additional continuing medical education after residency and changes in the medical school and residency curricula. In the meantime, we need more intervention studies to document the benefits of reducing exposures and innovations to bring the cost and feasibility of measuring chemical exposure into a range that would realistically allow for widespread use.

We also need to change the health professional mindset. Even within the community of people who focus on prevention, we have a long way to go to get chemical exposures, especially EDCs, into the conversation. Right now, in the global health world, the notion of "noncommunicable diseases" is being bandied about as a strategic priority. We've made so much progress when it comes to infectious conditions, especially HIV, tuberculosis, and malaria. Those challenges are by no means over yet. With that said, obesity is epidemic across developed and developing countries, followed by diabetes. All across the world, health ministries are reckoning with increases in male and female reproductive conditions, cancers, and neurodevelopmental disabilities[388] that could throw their budgets into disarray and threaten the progress we have made with life expectancy.

Look at the conversation around obesity and diabetes, however, and you see a singular focus on changing behaviors like diet and physical activity. Don't get me wrong—these are crucial factors in these epidemics. Also, throw in poverty, education, unemployment, and neighborhoods as insidious accelerators that contribute. Collectively, these are described as the social determinants of health. I have deep respect for the efforts to attack these problems. They, too, are environmental factors. Where I take issue, however, is the narrow definition of environment. EDCs are not even in the latest WHO report on noncommunicable diseases even though the WHO has acknowledged EDCs as an emerging public health issue![389]

Take a deeper look at the data on diet and physical activity interventions and their impact on obesity. The Cochrane Collaboration is a not-for-profit, highly respected international organization. Cochrane is so highly regarded because it actively digests and rigorously reviews the scientific literature so that policymakers and others can make informed decisions. In 2011 it reviewed 55 childhood obesity prevention studies that targeted 6- to 12-year-olds. There was strong evidence to support beneficial effects, but the broad array of interventions tried severely limited the interpretation. It was hard to know what exactly worked because some of the interventions were in homes, while others were in schools. Sometimes teachers received the school-based interventions, while in other cases the children were the focus.[390] And most of the studies that intervened to prevent obesity at the individual level didn't evaluate whether the effects were sustained into the long term. We all know how hard it is to maintain a healthy diet and physically active lifestyle. It can be very easy to fall off the proverbial wagon.

There's one other big factor to keep in mind. These interventions were resource intensive and costly. Even if you found the interventions that were consistently best and applied them, compare the costs for a health ministry to implement them among individuals in the population as opposed to the costs of regulating a chemical obesogen that may be produced by only a handful of manufacturers. Yes, there are costs to ensure adherence to the rule, but most of those costs are borne by the manufacturer and perhaps passed on to the consumer.

Am I saying we substitute a lifestyle approach to obesity

with a focus on reducing EDC exposures? Of course not. We can pursue both types of interventions. My point is simply: Given the urgency of the epidemic, why leave options off the table?

Social determinants as well as chemical exposures are part of a newer, broader mindset that may help us tackle all these factors together in a way that is much more productive. Christopher Wild, who directs the International Agency for Research on Cancer, coined the notion of the *exposome* in 2005. It's a broad notion that is inclusive of all types of environmental exposures (chemical, social, and physical, just to name a few). An underlying theme is that molecular techniques that we've already talked about in the book—epigenomics, metabolomics, measurements of chemical exposure in urine and blood, for example—can be used to measure the exposome and do everything from studying it to preventing and treating diseases that stem from adverse exposures.[391]

While there is no single definition of the social determinants of health, perhaps it's time to recalibrate to the broader scientific reality. If we don't shift our mindset, we will lag in our ability to address emerging challenges—and EDCs are one of them.

WE ARE THE CHANGE WE SEEK

The change we need to address EDCs is broad in scale—ideally, policy change would address it all. But we will need everyone to rise up, speak out, and act in our homes and our workplaces, with our family, friends, and colleagues. As patients, we must be empowered to ask about EDCs. As doctors and scientists, we all need to be accountable and accept that we may have to evolve our approach and mindset to embrace new scientific realities and challenges.

You've seen in this book multiple references to climate change. Of course, EDCs will be a secondary consideration if our planet becomes uninhabitable. But let's assume we can tackle climate change by reducing our carbon footprint. Even if we succeed, we may have contaminated ourselves, wildlife, and the broader ecosystem to the point that we may undermine the health gains we have made through better medical care and new technology.

I tell the undergraduates I teach at NYU that they are the best hope for our future. Millennials and Generation Z are more aware than previous generations of the environmental threats facing humanity and the planet. And I am not dismissing the role of the Greatest Generation, to use Tom Brokaw's description, in fixing these challenges. The approach I have found most effective with both newer and older generations, as well as my own, is to present the data I have and start a conversation about an issue like EDCs.

Some scientists feel an aversion to communicating a perspective about their own data, couching it with phrases such as "further research is needed." As you've seen in this book, much further research is needed to sort out the unknowns on EDCs, but I have a responsibility to report a fire when I see it. Endocrine-disrupting chemicals are the second greatest environmental challenge of our time. We won't tackle EDCs unless we have a broad-based conversation about chemicals, how they are tested for safety and used, and how we reckon with the constant reality that we will always be lagging in data. Yet we are sicker, fatter, and much poorer because of preventable EDC exposures.

So your job after reading this book is to take this message to others who haven't heard it yet. You may face resistance, debate, and disagreement. But every person you discuss EDCs with will have a family member with a chronic disease that may be due to these preventable exposures. They may come away wanting nothing to do with the issue. But they started knowing nothing about it in the first place.

When you do this, know that you're not alone. This book is based on the work of dozens of scientists who have labored for decades, navigating all sorts of challenges to protect the public. They have my gratitude and deepest respect. To borrow from Sir Isaac Newton, I stand on the shoulders of giants.

ACKNOWLEDGMENTS

This book would simply not have been possible without the stalwart support of my wife, Caitlin Aptowicz Trasande. Caitlin has been the rock of our family, and always encouraged me to go after this book and all of my other dreams. She held down the fort at home during the intense travel as the Endocrine Disrupting Chemicals (EDC) Disease Burden Working Group was deliberating in 2013 and 2014. Nurture matters, and our two boys Camilo and Ramiro are extremely lucky. I don't know how Caitlin does it all.

I also was lucky to marry into a family of writers, and it was through the partnership of my sister-in-law Cristin O'Keefe Aptowicz and her husband, Ernest Cline, that this book was born. I met Yfat Reiss Gendell of Foundry Literary + Media at the launch of Cristin's book *Dr. Mutter's Marvels: A True Tale of Intrigue and Innovation at the Dawn of Modern Medicine*. Yfat's special in so many ways—I realized only later that she'd written books on the endocrine system with her dad. She immediately grasped the science at a profound level, and was a master sherpa for me through the completely foreign universe of publishing.

Cristin and Ernie were amazing sounding boards at multiple points in the project. Writing a book is very much like a marathon, and I was very humbled to have two *New York Times*–best-selling authors to sense-check and bounce ideas.

As marathons go, book writing is very hilly, much like the Austin Marathon. Their emotional support was crucial to the final push for this book, which went much better than the final 11th Street hill when I ran Austin in 2017.

Yfat also introduced me to my partner in this project, the great Billie Fitzpatrick. As much as I tried not to be such a scientist in telling this story, Billie patiently guided me as we translated the science into accessible English. Billie deftly kept the references to scientific studies in the text while keeping the tone lively. It's easy to get depressed or distracted by the panoply of EDCs and effects, but I am upbeat about the present and future, and Billie made sure my voice and its optimism shone through.

I'm grateful to my thoughtful and passionate editor Deb Brody at Houghton Mifflin Harcourt, along with the tremendous support she received from manuscript editor Rebecca Springer, publisher Bruce Nicols, president Ellen Archer, managing editor Marina Padakis, production manager Tom Hyland, interior designer Chloe Foster and production coordinator Margaret Rosewitz, jacket designer Martha Kennedy, SVP of publicity Lori Glazer and publicity manager Sari Kamin, director of marketing Brianna Yamashita and senior marketing manager Brooke Bornemon, SVP of sales Maire Gorman, and editorial associate Olivia Bartz.

I'm also grateful to my audio publishing team at Audible Studios, including the Content Acquisition, Marketing, and Production teams, for their early support of this topic and my treatment of it. This book found a home because of their recognition that this subject needed a broad soapbox.

In addition, my thanks goes to the Foundry Literary + Media team, including Yfat's associate Jessica Felleman and assistant Anna Strzempko, foreign rights director Michael Nardullo and associate Heidi Gall, contracts director Deirdre Smerillo and associates Halyley Burdett and Melissa Moorehead, the filmed entertainment team, including director Richie Kern and associate Molly Gendell, and, importantly, their controller Sara DeNobrega and associates Collette Grecco and Sarah Lewis.

Many of my colleagues were extremely gracious to review early drafts of this book. These included: Jerry Heindel, Loretta Doan, Pete Myers, Kurunthachalam Kannan, Tom Zoeller, Barbara Demeneix, Sheela Sathyanarayana, Akhgar Ghassabian, Linda Kahn, and Pat Hunt.

Public health is also very much a team sport. I've been extremely privileged to work with a large group of wonderful colleagues who have been generous in so many ways that set the stage for this book. I will apologize now for any omissions, assuming there will be some despite my best efforts. First and foremost, Jerry Heindel was not only warm and encouraging of my own studies of EDCs, but a great partner in coordinating a wonderful Steering Committee for the EDC Disease Burden Working Group. Pete Myers also played a crucial role in bringing a world-class group of researchers together and helping identify funding for researchers to travel to meetings where most of the deliberations were completed. This book builds upon his beautiful description of EDC effects in *Our Stolen Future* with Theo Colborn and Diane Dumanowski.

Jerry and Pete are but two of the giants in the field of EDCs on whose shoulders this book stands. The other members of the Steering Committee were: Tom Zoeller, Ulla Hass, Andreas Kortenkamp, Philippe Grandjean, Joe DiGangi, Martine Bellanger, Russ Hauser, Juliette Legler and Niels Skakkebæk. The other members of the Working Group were Ana Soto, Paul A. Fowler, Patricia Hunt, Ruthann Rudel, Barbara Cohn, Frederic Bois, Soeren Ziebe, Sheela Sathyanarayana, Germaine Buck-Louis, Jorma Toppari, Anders Juul, Ulla Hass, Bruce Blumberg, Miquel Porta, Eva Govarts, and Barbara Demeneix.

The EDC Disease Burden Working Group also benefited from intellectual support from Annette Prüss-Ustun, Roberto Bertollini, and David Tordrup. Charles Persoz, Robert Barouki, and Marion Le Gal of the French National Alliance for Life Sciences and Health and Barbara Demeneix, Lindsey Marshall, Bilal Mughal, and Bolaji Seffou of the UMR 7221 Paris warmly hosted various meetings for the Working Group's first meeting in Paris in 2013. The Endocrine Society, the John Merck Fund, and the Oak Foundation also provided support for travel to the first and later meetings. The Ralph S. French Charitable Foundation provided additional support for the US disease burden and cost work.

I'd be remiss if I didn't communicate my profound gratitude to my many research collaborators. I won't even try to list them all, because I will most certainly leave someone out. Most especially, I will never be able to communicate my gratitude for the stalwart support of Teresa Attina, a research scientist who was my first hire at NYU back in 2011. As of this writing, Teresa and I have 19 publications in the scientific literature, including the manuscript in *Lancet Diabetes and Endocrinology* that documented $340 billion in EDC disease costs that we describe in this book. But that doesn't do justice to describing the profound support she has provided to the work that formed the basis of this book.

I was also extremely lucky to meet Jan Blustein very early in my time at NYU. Jan is a polymath with great emotional intelligence to boot. Countless times I've been able to lean on Jan for statistical support, writing advice, strategy navigating academic politics, you name it. Sheela Sathyanarayana, a fellow pediatrician whose work is also highlighted in this book, has been a joy to work with over the years. Vincent Jaddoe and his team in the Generation R Group at Erasmus University Medical Center have proven to be wonderful partners in looking at environmental influences on childhood obesity, and we are only getting started showing the fruits of our wonderful projects together.

New York University School of Medicine has proven to be a wonderful home for growing my research in this field. Vice Dean Dafna Bar-Sagi and my chair Katie Manno have long been champions of my work, providing resources and other support all along the way. Within the Department of Pediatrics, special thanks also go to Maria Ivanova, Ann Margaret McAdams, Brittany Leach, and Raymond Campbell. Marc Gourevitch and Max Costa direct the Departments of Population Health and Environmental Medicine and served on my mentorship committee, as did the great Benard Dreyer, George Thurston, and Adina Kalet. Arthur Fierman remains an equally enthusiastic colleague and friend after moving from the Division of General Pediatrics, which he led before establishing and directing my own Division of Environmental Pediatrics.

My own growing team in Environmental Pediatrics has been simply superb. Mrudula Naidu, Makhethe Mpoti, Tony Koshy, Joe Gilbert, and Garry Alcedo kept many of my projects rolling while I added book writing to my juggle. Akhgar Ghassabian, Linda Kahn, and Abby Gaylord are wonderful colleagues. I didn't bring Linda on board as a postdoctoral fellow because of her editorial experience in the publishing industry, though she was another great sounding board at various phases of this project. I'm also grateful to all the research staff that I've worked with in all of my projects over the years, beginning with studies of World Trade Center exposures in children back in 2013.

There's hardly a day when I don't think about the sacrifices made and challenges overcome by my parents, Dolores and Leonardo Trasande (Sr.), who arrived in the United States not long before I was born. Families like mine are further proof that the foundation of our country's greatness is its immigrants. Both my sister, Nancy, and I have tried to give back to our community through our legal and medical work (respectively) with low-income and largely Latino families. That mindset of giving back was inculcated by my parents from a very early age, and for that I am grateful.

Finally, my greatest gratitude is to the families whom I've had the privilege to serve throughout my career. They opened my eyes to the dangers of EDCs. I can only hope to give their stories, partially described in the vignettes in this book, the voice they deserve. Illnesses and costs can motivate, but each patient with a condition due to EDCs has her or his own story to tell. There are many other such stories that could not fit within the limits of this book. Thank you all for inspiring me to speak out about the need for clinicians, researchers, and policymakers to address one of the great public health crises of our time.

REFERENCES

INTRODUCTION

1. Trasande L, Zoeller RT, Hass U, et al. Estimating burden and disease costs of exposure to endocrine-disrupting chemicals in the European Union. *Journal of Clinical Endocrinology and Metabolism.* 2015:jc20144324.

2. Trasande L, Zoeller RT, Hass U, et al. Burden of disease and costs of exposure to endocrine disrupting chemicals in the European Union: an updated analysis. *Andrology.* 2016.

1. WHAT'S GOING ON?

3. Hinman AR, Orenstein WA, Schuchat A. Vaccine-preventable diseases, immunizations, and MMWR—1961–2011. *Morbidity and Mortality Weekly Reports.* 2011;60(4):49–57.

4. @CDCgov. Prevalence of autism spectrum disorder among children aged 8 years—autism and developmental disabilities monitoring network, 11 sites, United States, 2014 | *Morbidity and Mortality Weekly Reports.* 2018.

5. Visser SN, Danielson ML, Bitsko RH, et al. Trends in the parent-report of health care provider-diagnosed and medicated attention-deficit/hyperactivity disorder: United States, 2003–2011. *Journal of the American Academy of Child & Adolescent Psychiatry.* 2014;53(1):34–46.e32.

6. National Academies of Sciences, Engineering, and Medicine. In Oria MP, Stallings VA, eds. *Finding a Path to Safety in Food*

Allergy: Assessment of the Global Burden, Causes, Prevention, Management, and Public Policy. Washington, DC: National Academies Press, 2016.

7. Though the European Chemicals Agency, United States Environmental Protection Agency, the Endocrine Society, and the World Health Organization all have subtle differences in their definition of an EDC, the common thread and focus of this manuscript is the role of synthetic chemicals disrupting hormonal functions and contributing to disease and disability. Naturally occurring compounds do also disrupt endocrine functions, though it should be noted that endocrine-related conditions have increased in lockstep with the increase in synthetic chemical use. Below are citations that provide more details about the definition of an EDC: https://www.epa .gov/endocrine-disruption/what-endocrine-disruption; http:// ec.europa.eu/environment/chemicals/endocrine/definitions/ endodis_en.htm; http://www.who.int/ceh/risks/cehemerging 2/en/; https://www.endocrine.org/topics/edc.

8. Diamanti-Kandarakis E, Bourguignon J-P, Giudice LC, et al. Endocrine-disrupting chemicals: An Endocrine Society scientific statement. *Endocrine Reviews.* 2009;30(4):293–342.

9. Bergman Å, Heindel JJ, Jobling S, Kidd KA, Zoeller RT, eds. Global assessment of state-of-the-science for endocrine disruptors. 2012; Available at http://www.who.int/ipcs/publications/ new_issues/endocrine_disruptors/en/ (Accessed October 6, 2014).

10. The focus of this book, began, as chance would have it, when I was advising the United Nations Environment Programme in Geneva in 2011. During a coffee break, I ran into Dr. Jerry Heindel, a highly respected leader in the field who was then working at the National Institute of Environmental Health Sciences and known also for his infamous Hawaiian shirts. Jerry was helping to lead a team of scientists to develop a report for the World Health Organization and the United Nations Environment Programme describing the public health threat posed by these chemicals. He asked me if I could speak with the group

for a few minutes about the general approach that might be taken to estimate the disease burden and costs of these chemicals. I had by then developed a strong reputation for doing this kind of work, documenting the costs associated with prenatal mercury exposure and air pollution effects on the lungs of children, for example. That said, I had just begun to do studies of the effects of endocrine disruptors and was humbled to offer my insights to an all-star group of scientists. I was a newly minted associate professor surrounded by senior professors from all over the world, people whose names I recognized from work I had read and cherished, and some of whom you will meet in this book. After about 30 minutes, I felt good about making a positive impression and came away impressed by the passion in the room for accurately communicating the real threat posed by chemicals that disrupt the hormones in our bodies. I was even more energized to push my research in this direction.

11. Di Renzo GC, Conry JA, Blake J, et al. International Federation of Gynecology and Obstetrics opinion on reproductive health impacts of exposure to toxic environmental chemicals. *International Journal of Gynecology and Obstetrics*. 2015 Dec;131(3):219–225.

12. Gore AC, Chappell VA, Fenton SE, et al. EDC-2: The Endocrine Society's second scientific statement on endocrine-disrupting chemicals. *Endocrine Review*. 2015:er20151010.

13. Trasande L, Shaffer RM, Sathyanarayana S; American Academy of Pediatrics Council on Environmental Health. Food Additives and Child Health. *Pediatrics*. 2018;142(2):e20181408.

14. Centers for Disease Control and Prevention. National report on human exposure to environmental chemicals, updated tables, March 2018; https://www.cdc.gov/exposurereport/.

15. Official Journal of the European Union. Regulation (EC) No 1223/2009 of the European Parliament and of the Council. http://eur-lex.europa.eu/legal-content/EN/TXT/PDF/?uri=C ELEX:32009R1223&from=EN.

16. Center for Food Safety and Applied Nutrition. Laws & Regulations—Prohibited & Restricted Ingredients. https://www

.fda.gov/Cosmetics/GuidanceRegulation/LawsRegulations/ucm127406.htm (Accessed June 12, 2018).

17. State of California Department of Consumer Affairs. Technical bulletin 117-2013, Bureau of Electronic & Appliance Repair, Home Furnishings and Thermal Insulation. 2013; http://www.bearhfti.ca.gov/laws/tb117_2013.pdf.

18. European Commission. Restriction of hazardous substances in electrical and electronic equipment—environment—European Commission. 2018; http://ec.europa.eu/environment/waste/rohs_eee/legis_rohs1_en.htm.

19. Attina TM, Hauser R, Sathyanarayana S, et al. Exposure to endocrine-disrupting chemicals in the USA: A population-based disease burden and cost analysis. *Lancet Diabetes & Endocrinology.* 2016;4(12):996–1003.

20. Baldwin KR, Phillips AL, Horman B, et al. Sex specific placental accumulation and behavioral effects of developmental Firemaster 550 exposure in Wistar rats. *Scientific Reports.* 2017;7(1):7118.

21. Belcher SM, Cookman CJ, Patisaul HB, Stapleton HM. In vitro assessment of human nuclear hormone receptor activity and cytotoxicity of the flame retardant mixture FM 550 and its triarylphosphate and brominated components. *Toxicology Letters.* 2014;228(2):93–102.

22. Rock KD, Horman B, Phillips AL, et al. EDC IMPACT: Molecular effects of developmental FM 550 exposure in Wistar rat placenta and fetal forebrain. *Endocrine Connections.* 2018;7(2):305–324.

23. Trasande L. Updating the Toxic Substances Control Act to protect human health. *JAMA.* 2016;315(15):1565–1566.

24. Trasande L. When enough data are not enough to enact policy: The failure to ban chlorpyrifos. *PLoS Biology.* 2017;15(12):e2003671.

25. Hill A. The environment and disease: Association or causation? *Proceedings of the Royal Society of Medicine.* 1965;58(5):295–300.

26. You may have read about the Flynn effect, the notion that IQ has increased consistently worldwide since the 1930s. James Flynn is a New Zealand researcher who has his own TED talk. Some will use this data to suggest that chemicals are not

a problem. That said, look more carefully and you find a different trend. There's been a halt in verbal IQ progress among teenagers since the 1990s. These are children who were born in the 1970s, as chemical exposures were increasing. In Britain, Denmark, and France, overall IQ has actually decreased 2 to 4 IQ points! See E. Dutton, R. Lynn. *Intelligence* 51 (2015) 67–70. And, for perspective here, it's hard to interpret population-wide trends, where many competing factors are going up and down at the same time. That's why we rely on population-based studies where we can measure effects, a person at a time. And, for lead, and as you will see for many other chemicals, the evidence is very strong in humans and in the laboratory.

27. Grosse SD, Matte TD, Schwartz J, Jackson RJ. Economic gains resulting from the reduction in children's exposure to lead in the United States. *Environmental Health Perspectives.* 2002;110(6):563–569.

28. Tsai PL, Hatfield TH. Global benefits of phasing out leaded fuel. *Journal of Environmental Health.* 2011;74(5):8–15.

29. Attina TM, Trasande L. Economic costs of childhood lead exposure in low- and middle-income countries. *Environmental Health Perspectives.* 2013;121(9):1097–1102.

2. FOLLOWING THE HORMONAL CLUES

30. Appel A. Delaney clause heads for the history books. *Nature.* 1995;376(6536):109.

31. Allen W. *The war on bugs.* White River Junction, VT: Chelsea Green Publishing, 2008.

32. Carson RL. *Silent Spring.* Boston: Houghton Mifflin Company, 1962.

33. Herbst AL, Ulfelder H, Poskanzer DC. Adenocarcinoma of the vagina. Association of maternal stilbestrol therapy with tumor appearance in young women. *New England Journal of Medicine.* 1971;284(15):878–881.

34. Hoover RN, Hyer M, Pfeiffer RM, et al. Adverse health outcomes in women exposed in utero to diethylstilbestrol. *New England Journal of Medicine.* 2011;365(14):1304–1314.

35. Troisi R, Hyer M, Hatch EE, et al. Medical conditions among adult offspring prenatally exposed to diethylstilbestrol. *Epidemiology.* 2013;24(3):430–438.

36. Mahalingaiah S, Hart JE, Wise LA, Terry KL, Boynton-Jarrett R, Missmer SA. Prenatal diethylstilbestrol exposure and risk of uterine leiomyomata in the Nurses' Health Study II. *American Journal of Epidemiology.* 2014;179(2):186–191.

37. Hatch EE, Troisi R, Palmer JR, et al. Prenatal diethylstilbestrol exposure and risk of obesity in adult women. *Journal of Developmental Origins of Health and Disease.* 2015;6(3):201–207.

38. Palmer JR, Herbst AL, Noller KL, et al. Urogenital abnormalities in men exposed to diethylstilbestrol in utero: A cohort study. *Environmental Health.* 2009;8:37.

39. Troisi R, Titus L, Hatch EE, et al. A prospective cohort study of prenatal diethylstilbestrol exposure and cardiovascular disease risk. *Journal of Clinical Endocrinology and Metabolism.* 2018;103(1):206–212.

40. Kalfa N, Paris F, Soyer-Gobillard MO, et al. Prevalence of hypospadias in grandsons of women exposed to diethylstilbestrol during pregnancy: A multigenerational national cohort study. *Fertility and Sterility.* 2011;95(8):2574–2577.

41. Vandenberg LN, Maffini MV, Sonnenschein C, et al. Bisphenol-A and the great divide: A review of controversies in the field of endocrine disruption. *Endocrine Reviews.* 2009;30(1):75–95.

42. The report by Herbst and his colleagues spawned an entire line of research at the National Institute of Environmental Health Sciences led by John Maclachlan and Retha Newbold. We started to realize that workers could suffer similar consequences from industrial spills, such as a chemical called Kepone used in the synthesis of a pesticide called Mirex. Regular scientific conferences on Estrogens in the Environment united developmental biologists, biochemists, and other scientists, producing the recognition of more and more chemicals that were estrogen active, despite having different chemical structures.

43. Carlsen E, Giwercman A, Keiding N, Skakkebæk NE. Evidence

for decreasing quality of semen during past 50 years. *BMJ.* 1992;305(6854):609–613.

44. Levine H, Jorgensen N, Martino-Andrade A, et al. Temporal trends in sperm count: a systematic review and meta-regression analysis. *Human Reproduction Update.* 2017;23(6):646–659.

45. Colborn T, Dumanoski D, Myers JP. *Our stolen future: Are we threatening our fertility, intelligence, and survival? A scientific detective story.* Boston, MA: Little, Brown; 1996.

46. Whyatt RM, Rauh V, Barr DB, et al. Prenatal insecticide exposures and birth weight and length among an urban minority cohort. *Environmental Health Perspectives.* 2004;112(10):1125–1132.

47. Rauh V, Arunajadai S, Horton M, et al. Seven-year neurodevelopmental scores and prenatal exposure to chlorpyrifos, a common agricultural pesticide. *Environmental Health Perspectives.* 2011;119:1196–1201.

48. Eskenazi B, Marks AR, Bradman A, et al. Organophosphate pesticide exposure and neurodevelopment in young Mexican-American children. *Environmental Health Perspectives.* 2007;115(5):792–798.

49. Engel SM, Wetmur J, Chen J, et al. Prenatal exposure to organophosphates, paraoxonase 1, and cognitive development in childhood. *Environmental Health Perspectives.* 2011;119(8):1182–1188.

50. Rauh VA, Perera FP, Horton MK, et al. Brain anomalies in children exposed prenatally to a common organophosphate pesticide. *Proceedings of the National Academy of Sciences of the United States of America.* 2012;109(20):7871–7876.

51. Rauh VA, Garcia WE, Whyatt RM, et al. Prenatal exposure to the organophosphate pesticide chlorpyrifos and childhood tremor. *Neurotoxicology.* 2015;51:80–86.

52. Hunt PA, Koehler KE, Susiarjo M, et al. Bisphenol A exposure causes meiotic aneuploidy in the female mouse. *Current Biology.* 2003;13(7):546–553.

53. de Vries A. Paracelsus. Sixteenth-century physician-scientist-philosopher. *New York State Journal of Medicine.* 1977;77(5):378–455.

54. Vandenberg L, Colborn T, Hayes T, et al. Hormones and endocrine-disrupting chemicals: Low-dose effects and nonmonotonic dose responses. *Endocrinology Review.* 2012;33(3):378–455.

55. Birnbaum LS. Environmental chemicals: Evaluating low-dose effects. *Environmental Health Perspectives.* 2012;120(4):A143–144.

56. Trasande L, Vandenberg LN, Bourguignon JP, et al. Peer-reviewed and unbiased research, rather than 'sound science', should be used to evaluate endocrine-disrupting chemicals. *Journal of Epidemiology and Community Health.* 2016;70(11):1051–1056.

57. Trasande L, Attina TM, Blustein J. Association between urinary bisphenol A concentration and obesity prevalence in children and adolescents. *JAMA.* 2012;308(11):1113–1121.

58. Fagin D. Toxicology: The learning curve. *Nature.* 2012;490 (7421):462–465.

59. vom Saal FS, Timms BG, Montano MM, Palanza P, Thayer KA, et al. Prostate enlargement in mice due to fetal exposure to low doses of estradiol or diethylstilbestrol and opposite effects at high doses. *Proceedings of the National Academy of Sciences of the United States of America.* 1997 Mar 4;94(5):2056–61.

60. Vandenberg L, Colborn T, Hayes T, et al.

61. Villar-Pazos S, Martinez-Pinna J, Castellano-Munoz M, et al. Molecular mechanisms involved in the non-monotonic effect of bisphenol-a on ca2+ entry in mouse pancreatic beta-cells. *Scientific Reports.* 2017;7(1):11770.

62. Tavernise S. FDA makes it official: BPA can't be used in baby bottles and cups. *New York Times.* July 17, 2012; www.nytimes .com/2012/07/18/science/fda-bans-bpa-from-baby-bottles-and-sippy-cups.html (Accessed July 18, 2012).

63. FDA's BPA ban: A small, late step in the right direction. U.S. PIRG. 2018; https://uspirg.org/blogs/blog/usp/fda%E2%80%99s-bpa-ban-small-late-step-right-direction.

64. Safer States. Adopted Policy. Available at http://www.safer states.com/bill-tracker/ (Accessed January 8, 2016).

65. Wheeler L. Boxer: Chemical bill came from industry. *The Hill.* March 17, 2015.

66. Trasande L. Updating the Toxic Substances Control Act to protect human health. *JAMA*. 2016;315(15):1565–1566.

67. In FDA studies, treatments were also given to some animals with chemicals known to induce the same changes that are being questioned for BPA. These "positive control" studies are a lot like the tests we sometimes conduct in patients who we want to be really sure have a negative tuberculosis test. Sometimes in immune-compromised patients, we will inject a small amount of some agent to which they are already immunized, such as mumps. If they react to the mumps but not to the PPD test we use for tuberculosis testing, you can be sure you didn't incorrectly diagnose someone as being free of tuberculosis. Similarly, to test effects of BPA on the brains of rodents, the FDA scientists gave one group BPA while at the same time giving another group a medication known to induce effects on the brain called propylthiouracil, which acts against thyroid hormone. In these FDA studies, some of the "positive controls" were negative, as were the BPA-exposed rodents. The results of the FDA studies were misinterpreted to say that BPA has no effect, when a perfectly reasonable and responsible interpretation of the negative results in the positive controls is not to interpret anything at all. We simply don't know what happened.

68. Teeguarden JG, Twaddle NC, Churchwell MI, et al. 24-hour human urine and serum profiles of bisphenol A: Evidence against sublingual absorption following ingestion in soup. *Toxicology and Applied Pharmacology*. 2015;288(2):131–142.

69. Abbasi J. Scientists call FDA statement on bisphenol A safety premature. *JAMA* 2018; https://jamanetwork.com/journals/jama/articlepdf/2675909/jama_Abbasi_2018_mn_180018.pdf.

70. *Planet in Peril*. Aired April 23, 2008. CNN.com—Transcripts. http://transcripts.cnn.com/TRANSCRIPTS/0804/23/acd.02.html.

71. Hauser R, Skakkebæk NE, Hass U, et al. Male reproductive disorders, diseases, and costs of exposure to endocrine-disrupting chemicals in the European Union. *Journal of Clinical Endocrinology and Metabolism*. 2015;100(4):1267–1277.

72. The World Health Organization has developed an approach

to evaluating human studies of air pollution to literally grade the evidence (see notes 71–76 for more information about the GRADE Working Group). The Danish Environmental Protection Agency and the National Toxicology Program are among the organizations that have spearheaded similar efforts to measure the quality of individual studies and come to some conclusion about the relative strength of the science linking chemicals to potential effects on human health.

73. Trasande L, Zoeller RT, Hass U, et al. Burden of disease and costs of exposure to endocrine disrupting chemicals in the European Union: an updated analysis. *Andrology.* 2016.

74. Bellanger M, Demeneix B, Grandjean P, et al. Neurobehavioral deficits, diseases and associated costs of exposure to endocrine disrupting chemicals in the European Union. *Journal of Clinical Endocrinology and Metabolism.* 2015:100(4):1256-66.

75. Hauser R, Skakkebæk NE, Hass U, et al. Male reproductive disorders, diseases, and costs of exposure to endocrine-disrupting chemicals in the European Union.

76. Hunt PA, Sathyanarayana S, Fowler PA, Trasande L. Female reproductive disorders, diseases, and costs of exposure to endocrine disrupting chemicals in the European Union. *Journal of Clinical Endocrinology and Metabolism.* 2016;101(4):1562–1570.

77. Legler J, Fletcher T, Govarts E, et al. Obesity, diabetes, and associated costs of exposure to endocrine-disrupting chemicals in the European Union. *Journal of Clinical Endocrinology and Metabolism.* 2015:100(4):1278-88.

78. Trasande L, Zoeller RT, Hass U, et al. Estimating burden and disease costs of exposure to endocrine disruptor chemicals in the European Union. *Journal of Clinical Endocrinology and Metabolism,* 2015;100(4):1245–1255.

3. THE ATTACK ON THE BRAIN AND NERVOUS SYSTEM

79. Nelson KB, Ellenberg JH. Predictors of epilepsy in children who have had febrile seizures. *New England Journal of Medicine.* 1976; 295:1029–1033.

80. @CDCgov. Prevalence of autism spectrum disorder among children aged 8 years—autism and developmental disabilities monitoring network, 11 sites, United States, 2014.

81. Visser SN, Danielson ML, Bitsko RH, et al.

82. Demeneix B. *Losing our minds: Chemical pollution and the mental health of future generations.* Oxford, UK: Oxford University Press, 2014.

83. Hinton CF, Harris KB, Borgfeld L, et al. Trends in incidence rates of congenital hypothyroidism related to select demographic factors: Data from the United States, California, Massachusetts, New York, and Texas. *Pediatrics.* 2010;125 Suppl 2:S37–47.

84. Bernal J. In memoriam: Gabriella Morreale de Escobar. *European Thyroid Journal.* 2018;7(2):109–110.

85. Haddow JE, Palomaki GE, Allan WC, et al. Maternal thyroid deficiency during pregnancy and subsequent neuropsychological development of the child. *New England Journal of Medicine.* 1999;341(8):549–555.

86. Peeters RP. Subclinical hypothyroidism. *New England Journal of Medicine.* 2017;376(26):2556–2565.

87. Korevaar TIM, Medici M, Visser TJ, et al. Thyroid disease in pregnancy: New insights in diagnosis and clinical management. *Nature Reviews Endocrinology.* 2017;13(10):610.

88. Bellanger M, Demeneix B, Grandjean P, et al.

89. Casey BM, Thom EA, Peaceman AM, et al. Treatment of subclinical hypothyroidism or hypothyroxinemia in pregnancy. *New England Journal of Medicine.* 2017;376:815–825.

90. United Nations Environment Programme (Stockholm Convention Secretariat). Stockholm Convention on Persistent Organic Pollutants. Available at http://chm.pops.int/default.aspx (Accessed December 8, 2010).

91. Jacobson JL, Jacobson SW. Intellectual impairment in children exposed to polychlorinated biphenyls in utero. *New England Journal of Medicine.* 1996;335(11):783.

92. Naveau E, Pinson A, Gerard A, et al. Alteration of rat fetal cerebral cortex development after prenatal exposure to polychlorinated biphenyls. *PLoS One.* 2014;9(3):e91903.

93. Gauger KJ, Kato Y, Haraguchi K, et al. Polychlorinated biphenyls (PCBs) exert thyroid hormone-like effects in the fetal rat brain but do not bind to thyroid hormone receptors. *Environmental Health Perspectives.* 2004;112(5):516–523.

94. Zoeller RT, Dowling ALS, Herzig CTA, et al. Thyroid hormone, brain development, and the environment. *Environmental Health Perspectives.* 2002;110(s3):355–361.

95. Hill A.

96. Summary of the Federal Insecticide, Fungicide, and Rodenticide Act. [Overviews and Factsheets]. 2018; https://www.epa.gov/laws-regulations/summary-federal-insecticide-fungicide-and-rodenticide-act.

97. Čolović MB, Krstić DZ, Lazarević-Pašti TD, Bondžić AM, Vasić VM. Acetylcholinesterase inhibitors: pharmacology and toxicology. *Current Neuropharmacology.* 2013;11(3):315–335.

98. De Angelis S, Tassinari R, Maranghi F, et al. Developmental exposure to chlorpyrifos induces alterations in thyroid and thyroid hormone levels without other toxicity signs in CD-1 mice. *Toxicological Sciences.* 2009;108(2):311–319.

99. Levin ED, Addy N, Baruah A, et al. Prenatal chlorpyrifos exposure in rats causes persistent behavioral alterations. *Neurotoxicology and Teratology.* 2002;24(6):733–741.

100. Berbel P, Auso E, Garcia-Velasco JV, et al. Role of thyroid hormones in the maturation and organisation of rat barrel cortex. *Neuroscience.* 2001;107(3):383–394.

101. This project would not have been possible without the support of two leading scientists who have played crucial roles in our understanding the effects of chemicals on thyroid hormone and brain development in animals. Tom Zoeller is a professor of biology at the University of Massachusetts at Amherst, and was a leader in the group that wrote the World Health Organization and United Nations Environment Programme report in 2012 that put endocrine disruption on the global public health map. Barbara Demeneix is a British biologist and endocrinologist at the Museum de la Histoire Naturelle, the Parisian version of the American Museum of Natural History. We were

joined in this project by French economist Martine Bellanger and Danish epidemiologist Phillipe Grandjean, who splits his time between the Harvard TH Chan School of Public Health and the Southern Denmark University.

102. Bellanger M, Demeneix B, Grandjean P, et al.

103. Rauh VA, Perera FP, Horton MK, et al.

104. Bellanger M, Demeneix B, Grandjean P, et al.

105. Grosse SD, Matte TD, Schwartz J, et al.

106. Revkin AC. E.P.A., citing risks to children, signs accord to limit insecticide. *New York Times.* June 9, 2000.

107. US EPA. Food Quality Protection Act (FQPA) of 1996. 1996; Available at http://www.epa.gov/pesticides/regulating/laws/fqpa/ (Accessed February 2, 2009).

108. US EPA. EPA administrator Pruitt denies petition to ban widely used pesticide. [Speeches, Testimony and Transcripts]. 2017; https://www.epa.gov/newsreleases/epa-administrator-pruitt-denies-petition-ban-widely-used-pesticide-0.

109. Seufert V, Ramankutty N, Foley JA. Comparing the yields of organic and conventional agriculture. *Nature.* 2012;485:229–232.

110. United Nations Human Rights Office of the High Commissioner. Special rapporteur on the right to food. 2017; http://www.ohchr.org/EN/Issues/Food/Pages/FoodIndex.aspx. (Accessed June 29, 2017).

111. Trasande L. When enough data are not enough to enact policy: The failure to ban chlorpyrifos.

112. Lu C, Toepel K, Irish R, et al. Organic diets significantly lower children's dietary exposure to organophosphorus pesticides. *Environmental Health Perspectives.* 2006;114(2):260–263.

113. Bradman A, Quiros-Alcala L, Castorina R, et al. Effect of organic diet intervention on pesticide exposures in young children living in low-income urban and agricultural communities. *Environmental Health Perspectives.* 2015;123(10):1086–1093.

114. Bellanger M, Demeneix B, Grandjean P, et al.

115. Herbstman JB, Sjödin A, Kurzon M, et al. Prenatal exposure to PBDEs and neurodevelopment. *Environmental Health Perspectives.* 2010;118(5):712–719.

116. Eskenazi B, Chevrier J, Rauch SA, et al. In utero and childhood polybrominated diphenyl ether (PBDE) exposures and neurodevelopment in the CHAMACOS study. *Environmental Health Perspectives.* 2013;121(2):257–262.

117. Chen A, Yolton K, Rauch SA, et al. Prenatal polybrominated diphenyl ether exposures and neurodevelopment in U.S. children through 5 years of age: The HOME study. *Environmental Health Perspectives.* 2014;122(8):856–862.

118. Gascon M, Vrijheid M, Martinez D, et al. Effects of pre and postnatal exposure to low levels of polybromodiphenyl ethers on neurodevelopment and thyroid hormone levels at 4 years of age. *Environment International.* 2011;37(3):605–611.

119. Gascon M, Vrijheid M, Martinez D, et al.

120. Attina TM, Hauser R, Sathyanarayana S, et al.

121. Attina TM, Malits J, Naidu M, Trasande L. Racial/Ethnic Disparities in Disease Burden and Costs Related to Exposure to Endocrine Disrupting Chemicals in the US: an Exploratory Analysis. J Clin Epidemiol. 2018 Dec 6. pii: S0895-4356(18)30451-7. doi: 10.1016/j.jclinepi.2018.11.024. [Epub ahead of print]

122. Gomis MI, Vestergren R, Borg D, et al. Comparing the toxic potency in vivo of long-chain perfluoroalkyl acids and fluorinated alternatives. *Environment International.* 2018;113:1–9.

123. Beekman M, Zweers P, Muller A, et al. Evaluation of substances used in the GenX technology by Chemours, Dordrecht—RIVM. 2016; https://www.rivm.nl/en/Documents_and_publications/Scientific/Reports/2016/december/Evaluation_of_substances_used_in_the_GenX_technology_by_Chemours_Dordrecht.

124. Lerner S. New Teflon toxin found in North Carolina drinking water. 2018; https://theintercept.com/2017/06/17/new-teflon-toxin-found-in-north-carolina-drinking-water/.

125. Kim S, Jung J, Lee I, et al. Thyroid disruption by triphenyl phosphate, an organophosphate flame retardant, in zebrafish (Danio rerio) embryos/larvae, and in GH3 and FRTL-5 cell lines. *Aquatic Toxicology.* 2015;160:188–196.

126. Chen A, Yolton K, Rauch SA, et al.

127. Gascon M, Vrijheid M, Martinez D, et al.

128. Marks AR, Harley K, Bradman A, et al. Organophosphate pesticide exposure and attention in young Mexican-American children: The CHAMACOS study. *Environmental Health Perspectives.* 2010;118(12):1768–1774.

129. McDonald MP, Wong R, Goldstein G, et al. Hyperactivity and learning deficits in transgenic mice bearing a human mutant thyroid hormone beta1 receptor gene. *Learning & Memory.* 1998;5(4):289–301.

130. Akaike M, Kato N, Ohno H, et al. Hyperactivity and spatial maze learning impairment of adult rats with temporary neonatal hypothyroidism. *Neurotoxicology and Teratology.* 1991;13(3):317–322.

131. Kiguchi M, Fujita S, Oki H, et al. Behavioural characterisation of rats exposed neonatally to bisphenol-A: Responses to a novel environment and to methylphenidate challenge in a putative model of attention-deficit hyperactivity disorder. *Journal of Neural Transmission (Vienna)* 2008;115(7):1079–1085.

132. Sazonova NA, DasBanerjee T, Middleton FA, et al. Transcriptome-wide gene expression in a rat model of attention deficit hyperactivity disorder symptoms: Rats developmentally exposed to polychlorinated biphenyls. *American Journal of Medical Genetics Part B, Neuropsychiatric Genetics.* 2011;156b(8):898–912.

133. Miodovnik A, Engel SM, Zhu C, et al. Endocrine disruptors and childhood social impairment. *Neurotoxicology.* 2011;32(2):261–267.

134. Braun JM, Kalkbrenner AE, Just AC, et al. Gestational exposure to endocrine-disrupting chemicals and reciprocal social, repetitive, and stereotypic behaviors in 4- and 5-year-old children: The HOME study. *Environmental Health Perspectives.* 2014;122(5):513–520.

4. METABOLIC MIX-UPS: OBESITY AND DIABETES

135. Hales CM, Fryar CD, Carroll MD, et al. Trends in obesity and severe obesity prevalence in US youth and adults by sex and age, 2007–2008 to 2015–2016. *JAMA.* 2018;319(16):1723–1725.

136. Cawley J, Meyerhoefer C. The medical care costs of obesity:

An instrumental variables approach. *Journal of Health Economics.* 2012;31(1):219–230.

137. Brown RE, Sharma AM, Ardern CI, et al. Secular differences in the association between caloric intake, macronutrient intake, and physical activity with obesity. *Obesity Research & Clinical Practice.* 2016;10(3):243–255.

138. Lustig RH. Fructose: It's "alcohol without the buzz." *Advances in Nutrition.* 2013;4(2):226–235.

139. Redline S, Tishler PV, Schluchter M, et al. Risk factors for sleep-disordered breathing in children. Associations with obesity, race, and respiratory problems. *American Journal of Respiratory and Critical Care Medicine.* 1999;159(5 Pt 1):1527–1532.

140. Gordon-Larsen P, Nelson MC, Page P, et al. Inequality in the built environment underlies key health disparities in physical activity and obesity. *Pediatrics.* 2006;117(2):417–424.

141. Liu G, Dhana K, Furtado JD, et al. Perfluoroalkyl substances and changes in body weight and resting metabolic rate in response to weight-loss diets: A prospective study. *PLoS Medicine.* 2018;15(2):e1002502.

142. Barker DJ, Osmond C. Infant mortality, childhood nutrition, and ischaemic heart disease in England and Wales. *Lancet.* 1986;1(8489):1077–1081.

143. Hales CN, Barker DJP. The thrifty phenotype hypothesis type 2 diabetes. *British Medical Bulletin.* 2001;60(1):5–20.

144. Vaag AA, Grunnet LG, Arora GP, Brøns C. The thrifty phenotype hypothesis revisited. *Diabetologia.* 2012;55(8):2085–2088.

145. Haugen AC, Schug TT, Collman G, et al. Evolution of DOHaD: The impact of environmental health sciences. *Journal of Developmental Origins of Health and Disease.* 2015;6(2):55–64.

146. Trasande L, Cronk C, Durkin M, et al. Environment, obesity and the National Children's Study. *Environmental Health Perspectives.* 2009;117(2):159–166.

147. Reardon S. NIH ends longitudinal children's study. *Nature.* December 12, 2014. doi:10.1038/nature.2014.16556

148. Mayer-Davis EJ, Lawrence JM, Dabelea D, et al. Incidence trends of type 1 and type 2 diabetes among youths, 2002–2012. *New England Journal of Medicine.* 2017;376(15):1419–1429.

149. Ruiz D, Becerra M, Jagai JS, et al. Disparities in environmental exposures to endocrine-disrupting chemicals and diabetes risk in vulnerable populations. *Diabetes Care.* 2018;41(1): 193–205.

150. Janesick A, Blumberg B. Obesogens, stem cells and the developmental programming of obesity. *International Journal of Andrology.* 2012;35(3):437–448.

151. Kirchner S, Kieu T, Chow C, et al. Prenatal exposure to the environmental obesogen tributyltin predisposes multipotent stem cells to become adipocytes. *Molecular Endocrinology.* 2010;24(3):526–539.

152. Sathyanarayana S. Phthalates and children's health. *Current Problems in Pediatric and Adolescent Health Care.* 2008;38(2):34–49.

153. Serrano SE, Braun J, Trasande L, et al. Phthalates and diet: A review of the food monitoring and epidemiology data. *Environmental Health.* 2014;13(1):43.

154. Ferguson KK, Loch-Caruso R, Meeker JD. Urinary phthalate metabolites in relation to biomarkers of inflammation and oxidative stress from NHANES 1999–2006. *Environmental Research.* 2011;111(5):718–726.

155. Ceriello A, Motz E. Is oxidative stress the pathogenic mechanism underlying insulin resistance, diabetes, and cardiovascular disease? The common soil hypothesis revisited. *Arteriosclerosis, Thrombosis, and Vascular Biology.* 2004;24(5):816–823.

156. Posnack NG, Lee NH, Brown R, et al. Gene expression profiling of DEHP-treated cardiomyocytes reveals potential causes of phthalate arrhythmogenicity. *Toxicology.* 2011;279(1–3):54–64.

157. Meeker JD, Ferguson KK. Urinary phthalate metabolites are associated with decreased serum testosterone in men, women, and children from NHANES 2011–2012. *Journal of Clinical Endocrinology & Metabolism.* 2014;99(11):4346–4352.

158. Pan G, Hanaoka T, Yoshimura M, et al. Decreased serum free testosterone in workers exposed to high levels of di-n-butyl phthalate (DBP) and di-2-ethylhexyl phthalate (DEHP): A cross-sectional study in China. *Environmental Health Perspectives.* 2006;114(11):1643–1648.

159. Holmboe SA, Skakkebæk NE, Juul A, et al. Individual testoster-

one decline and future mortality risk in men. *European Journal of Endocrinology.* 2018;178(1):123–130.

160. Kelly DM, Jones TH. Testosterone: A vascular hormone in health and disease. *Journal of Endocrinology.* 2013;217(3):R47–71.

161. Oskui PM, French WJ, Herring MJ, et al. Testosterone and the cardiovascular system: A comprehensive review of the clinical literature. *Journal of the American Heart Association.* 2013;2(6):e000272.

162. Morgentaler A, Traish A, Kacker R. Deaths and cardiovascular events in men receiving testosterone. *JAMA.* 2018;311(9):961–962.

163. Miner M, Morgentaler A, Khera M, et al. The state of testosterone therapy since the FDA's 2015 labeling changes: Indications and cardiovascular risk. *Clinical Endocrinology (Oxford).* 2018.

164. Vigen R, O'Donnell CI, Baron AE, et al. Association of testosterone therapy with mortality, myocardial infarction, and stroke in men with low testosterone levels. *JAMA.* 2013;310(17):1829–1836.

165. Yeap BB. Testosterone and ill-health in aging men. *Nature Clinical Practice Endocrinology and Metabolism.* 2009;5(2):113–121.

166. Kelly DM, Jones TH.

167. Legler J, Fletcher T, Govarts E, et al. Obesity, diabetes, and associated costs of exposure to endocrine-disrupting chemicals in the European Union. *Journal of Clinical Endocrinology and Metabolism.* 2015;100(4):1278–1288.
The experts that examined EDCs as obesogens and cardiovascular risks were led by Juliette Legler, a superb and energetic researcher now based at Utrecht University in the Netherlands and one of the few who has been able to bridge the laboratory work, or toxicology, with the human studies, or epidemiology. Miquel Porta, a Catalonian physician-scientist, kept us entertained through tough deliberations with his wry sense of humor. Tony Fletcher, an epidemiologist at the London School of Tropical Hygiene and Medicine known for his studies of PFOA and other "long-chain" PFASs in a large community in West Virginia living near a plant that contaminated the water supply, warmed to the project over time and helped us maintain a very careful and conservative approach. Eva Govarts, a Belgian scientist,

was a superb addition, helping us take the available data on relevant exposures and ensuring we had the best data possible on which to make estimates. After a "lekker" canal tour at the end of our second meeting, we liked working together and wanted to keep going, but had to get back to our day jobs, as we were not getting paid for this project.

168. Philips EM, Jaddoe VWV, Trasande L. Effects of early exposure to phthalates and bisphenols on cardiometabolic outcomes in pregnancy and childhood. *Reproductive Toxicology*. 2017;68:105–118.

169. Song Y, Hauser R, Hu FB, et al. Urinary concentrations of bisphenol A and phthalate metabolites and weight change: A prospective investigation in US women. *International Journal of Obesity (London)*. 2014;38(12):1532–1537.

170. Lind PM, Roos V, Ronn M, et al. Serum concentrations of phthalate metabolites are related to abdominal fat distribution two years later in elderly women. *Environmental Health*. 2012;11(1):21.

171. Hill A.

172. Stahlhut RW, van Wijngaarden E, Dye TD, et al. Concentrations of urinary phthalate metabolites are associated with increased waist circumference and insulin resistance in adult U.S. males. *Environmental Health Perspectives*. 2007;115(6):876–882.

173. Trasande L, Attina TM, Sathyanarayana S, et al. Race/ethnicity-specific associations of urinary phthalates with childhood body mass in a nationally representative sample. *Environmental Health Perspectives*. 2013;121(4):501–506.

174. Trasande L, Spanier AJ, Sathyanarayana S, et al. Urinary phthalates and increased insulin resistance in adolescents. *Pediatrics*. 2013;132(3):e646–655.

175. Trasande L, Sathyanarayana S, Spanier AJ, et al. Urinary phthalates are associated with higher blood pressure in childhood. *Journal of Pediatrics*. 2013;163(3):747–753.e741.

176. Attina TM, Trasande L. Association of exposure to di-2-ethylhexylphthalate replacements with increased insulin resistance in adolescents from NHANES 2009–2012. *Journal of Clinical Endocrinology and Metabolism*. 2015;100(7):2640–2650.

177. Trasande L, Attina TM. Association of exposure to di-2-ethyl-hexylphthalate replacements with increased blood pressure in children and adolescents. *Hypertension.* 2015;66(2):301–308.

178. GBD Risk Factors Collaborators, Forouzanfar MH, Alexander L, et al. Global, regional, and national comparative risk assessment of 79 behavioural, environmental and occupational, and metabolic risks or clusters of risks in 188 countries, 1990–2013: A systematic analysis for the Global Burden of Disease Study 2013. *Lancet.* 2015;386(10010):2287–2323.

179. Schultz TW, Sinks GD. Xenoestrogenic gene expression: Structural features of active polycyclic aromatic hydrocarbons. *Environmental Toxicology and Chemistry.* 2002;21(4):783–786.

180. Vinggaard AM, Hnida C, Larsen JC. Environmental polycyclic aromatic hydrocarbons affect androgen receptor activation in vitro. *Toxicology.* 2000;145(2–3):173–183.

181. Sun H, Shen O-X, Xu X-L, et al. Carbaryl, 1-naphthol and 2-naphthol inhibit the beta-1 thyroid hormone receptor-mediated transcription in vitro. *Toxicology.* 2008;249(2–3):238–242.

182. Kim JH, Yamaguchi K, Lee SH, et al. Evaluation of polycyclic aromatic hydrocarbons in the activation of early growth response-1 and peroxisome proliferator activated receptors. *Toxicological Sciences.* 2005;85(1):585–593.

183. Rundle A, Hoepner L, Hassoun A, et al. Association of childhood obesity with maternal exposure to ambient air polycyclic aromatic hydrocarbons during pregnancy. *American Journal of Epidemiology.* 2012;175(11):1163–1172.

184. Wolf K, Popp A, Schneider A, et al. Association between long-term exposure to air pollution and biomarkers related to insulin resistance, subclinical inflammation, and adipokines. *Diabetes.* 2016;65(11):3314–3326.

185. Rajagopalan S, Brook RD. Air pollution and type 2 diabetes: Mechanistic insights. *Diabetes.* 2012;61(12):3037–3045.

186. Hu FB, Satija A, Manson JE. Curbing the diabetes pandemic: The need for global policy solutions. *JAMA.* 2015;313(23):2319–2320.

187. Masuno H, Kidani T, Sekiya K, et al. Bisphenol A in combina-

tion with insulin can accelerate the conversion of 3T3-L1 fibroblasts to adipocytes. *Journal of Lipid Research.* 2002;43(5):676–684.

188. Sakurai K, Kawazuma M, Adachi T, et al. Bisphenol A affects glucose transport in mouse 3T3-F442A adipocytes. *British Journal of Pharmacology.* 2004;141(2):209–214.

189. Hugo ER, Brandebourg TD, Woo JG, et al. Bisphenol A at environmentally relevant doses inhibits adiponectin release from human adipose tissue explants and adipocytes. *Environmental Health Perspectives.* 2008;116(12):1642–1647.

190. Schecter A, Malik N, Haffner D, et al. Bisphenol A (BPA) in U.S. food. *Environmental Science & Technology.* 2010;44(24):9425–9430.

191. Morgan MK, Jones PA, Calafat AM, et al. Assessing the quantitative relationships between preschool children's exposures to bisphenol A by route and urinary biomonitoring. *Environmental Science & Technology.* 2011;45(12):5309–5316.

192. Schecter A, Malik N, Haffner D, et al.

193. Wassenaar PNH, Trasande L, Legler J. Systematic review and meta-analysis of early-life exposure to bisphenol A and obesity-related outcomes in rodents. *Environmental Health Perspectives.* 2017;125(10):106001.

194. Hoepner LA, Whyatt RM, Widen EM, et al. Bisphenol A and adiposity in an inner-city birth cohort. *Environmental Health Perspectives.* 2016;124(10):1644–1650.

195. Valvi D, Casas M, Mendez MA, et al. Prenatal bisphenol A urine concentrations and early rapid growth and overweight risk in the offspring. *Epidemiology.* 2013;24(6):791–799.

196. Harley KG, Aguilar Schall R, Chevrier J, et al. Prenatal and postnatal bisphenol A exposure and body mass index in childhood in the CHAMACOS cohort. *Environmental Health Perspectives.* 2013;121(4):514–520.

197. Braun JM, Lanphear BP, Calafat AM, et al. Early-life bisphenol A exposure and child body mass index: A prospective cohort study. *Environmental Health Perspectives.* 2014;122(11):1239–1245.

198. Stahlhut RW, Welshons WV, Swan SH. Bisphenol A data in NHANES suggest longer than expected half-life, substantial

nonfood exposure, or both. *Environmental Health Perspectives.* 2009;117(5):784–789.

199. Snijder CA, Heederik D, Pierik FH, et al. Fetal growth and prenatal exposure to bisphenol A: The generation R study. *Environmental Health Perspectives.* 2013;121(3):393–398.

200. Hoepner LA, Whyatt RM, Widen EM, et al.

201. Melzer D, Rice NE, Lewis C, et al. Association of urinary bisphenol A concentration with heart disease: Evidence from NHANES 2003/06. *PLoS One.* 2010;5(1):e8673.

202. Melzer D, Gates P, Osborn NJ, et al. Urinary bisphenol A concentration and angiography-defined coronary artery stenosis. *PLoS One.* 2012;7(8):e43378.

203. Melzer D, Osborne NJ, Henley WE, et al. Urinary bisphenol A concentration and risk of future coronary artery disease in apparently healthy men and women. *Circulation.* 2012;125(12):1482–1490.

204. Trasande L. Further limiting bisphenol A in food uses could provide health and economic benefits. *Health Affairs (Millwood).* 2014;33(2):316–323.

205. Harley KG, Kogut K, Madrigal DS, et al. Reducing phthalate, paraben, and phenol exposure from personal care products in adolescent girls: Findings from the HERMOSA Intervention Study. *Environmental Health Perspectives.* 2016;124(10):1600–1607.

206. Rudel RA, Gray JM, Engel CL, et al. Food packaging and bisphenol A and bis(2-ethylhexyl) phthalate exposure: Findings from a dietary intervention. *Environmental Health Perspectives.* 2011;119(7):914–920.

207. Serrano SE, Braun J, Trasande L, et al.

208. Ax E, Lampa E, Lind L, et al. Circulating levels of environmental contaminants are associated with dietary patterns in older adults. *Environment International.* 2015;75:93–102.

209. Sjogren P, Montse R, Lampa E, et al. Circulating levels of perfluoroalkyl substances are associated with dietary patterns — A cross sectional study in elderly Swedish men and women. *Environmental Research.* 2016;150:59–65.

210. Dolinoy DC, Huang D, Jirtle RL. Maternal nutrient supplemen-

tation counteracts bisphenol A-induced DNA hypomethylation in early development. *Proceedings of the National Academy of Sciences of the United States of America.* 2007;104(32):13056–13061.

211. Doerr A. Global metabolomics. *Nature Methods.* 2016;14(1):32.

212. Trasande L. Further limiting bisphenol A in food uses could provide health and economic benefits.

213. Kuruto-Niwa R, Nozawa R, Miyakoshi T, et al. Estrogenic activity of alkylphenols, bisphenol S, and their chlorinated derivatives using a GFP expression system. *Environmental Toxicology and Pharmacology.* 2005;19(1):121–130.

214. Chen MY, Ike M, Fujita M. Acute toxicity, mutagenicity, and estrogenicity of bisphenol-A and other bisphenols. *Environmental Toxicology.* 2002;17(1):80–86.

215. Yoshihara Si, Mizutare T, Makishima M, et al. Potent estrogenic metabolites of bisphenol A and bisphenol B formed by rat liver S9 fraction: Their structures and estrogenic potency. *Toxicological Sciences.* 2004;78(1):50–59.

216. Okuda K, Fukuuchi T, Takiguchi M, et al. Novel pathway of metabolic activation of bisphenol A-related compounds for estrogenic activity. *Drug Metabolism and Disposition.* 2011;39(9):1696–1703.

217. Audebert M, Dolo L, Perdu E, et al. Use of the γH2AX assay for assessing the genotoxicity of bisphenol A and bisphenol F in human cell lines. *Archives of Toxicology.* 2011;85(11):1463–1473.

218. Danzl E, Sei K, Soda S, et al. Biodegradation of bisphenol A, bisphenol F and bisphenol S in seawater. *International Journal of Environmental Research and Public Health.* 2009;6(4):1472–1484.

219. Ike M, Chen MY, Danzl E, et al. Biodegradation of a variety of bisphenols under aerobic and anaerobic conditions. *Water Science and Technology.* 2006;53(6):153–159.

5. A REAL-LIFE CHILDREN OF MEN?

220. Carlsen E, Giwercman A, Keiding N, Skakkebæk NE.

221. Skakkebæk NE, Rajpert-De Meyts E, Main KM. Testicular dysgenesis syndrome: An increasingly common developmen-

tal disorder with environmental aspects. *Human Reproduction.* 2001;16(5):972–978.

222. Bergman A, Heindel JJ, Jobling S, et al. State of the science of endocrine disrupting chemicals 2012. United Nations Environment Programme and World Health Organization; 2013.

223. Paulozzi LJ, Erickson JD, Jackson RJ. Hypospadias trends in two US surveillance systems. *Pediatrics.* 1997;100(5):831–834.

224. Bergman A, Heindel JJ, Jobling S, et al.

225. Holmes L, Jr., Escalante C, Garrison O, et al. Testicular cancer incidence trends in the USA (1975–2004): Plateau or shifting racial paradigm? *Public Health.* 2008;122(9):862–872.

226. Wang Z, McGlynn KA, Meyts ER-D, et al. Meta-analysis of five genome-wide association studies identifies multiple new loci associated with testicular germ cell tumor. *Nature Genetics.* 2017;49(7):1141.

227. McGlynn KA, Trabert B. Adolescent and adult risk factors for testicular cancer. *Nature Reviews Urology.* 2012;9(6):339–349.

228. Bay K, Main KM, Toppari J, et al. Testicular descent: INSL3, testosterone, genes and the intrauterine milieu. *Nature Reviews Urology.* 2011;8(4):187–196.

229. Trasande L. Clinical awareness of occupation-related toxic exposure. *Virtual Mentor.* 2006;8:723–728. Available at http://www.ama-assn.org/ama/pub/category/16932.html.

230. McGlynn KA, Trabert B.

231. Ibid.

232. McGlynn KA, Quraishi SM, Graubard BI, et al. Persistent organochlorine pesticides and risk of testicular germ cell tumors. *Journal of the National Cancer Institute.* 2008;100(9):663–671.

233. Purdue MP, Engel LS, Langseth H, et al. Prediagnostic serum concentrations of organochlorine compounds and risk of testicular germ cell tumors. *Environmental Health Perspectives.* 2009;117(10):1514–1519.

234. Hardell L, Bavel B, Lindstrom G, et al. In utero exposure to persistent organic pollutants in relation to testicular cancer risk. *International Journal of Andrology.* 2006;29(1):228–234.

235. Stoker TE, Cooper RL, Lambright CS, et al. In vivo and in vi-

tro anti-androgenic effects of DE-71, a commercial polybrominated diphenyl ether (PBDE) mixture. *Toxicology and Applied Pharmacology.* 2005;207(1):78–88.

236. McGlynn KA, Trabert B.

237. Testicular Cancer—Cancer Stat Facts. 2018; https://seer.cancer.gov/statfacts/html/testis.html.

238. A short walk down Shattuck Street in Boston would take you from the hospital where I completed my surgical rotation to the Harvard T.H. Chan School of Public Health, where my future colleague Russ Hauser had just completed his doctoral training in public health. I would first get to know Russ and Niels through a panel of experts we organized on EDCs and male reproductive health. We were joined by Jorma Toppari, a Finnish pediatrician-scientist; Andreas Kortenkamp, a German toxicologist working at Brunel University in London; Ulla Hass, a Danish reproductive toxicologist; and two colleagues of Niels, Anders Juul and Anna Maria Andersson.

239. Virtanen HE, Toppari J. Epidemiology and pathogenesis of cryptorchidism. *Human Reproduction Update.* 2008;14(1):49–58.

240. Hauser R, Skakkebæk NE, Hass U, et al.

241. Goodyer CG, Poon S, Aleksa K, Hou L, Aterhortua V, et al. A case-control study of maternal polybrominated diphenyl ether (PBDE) exposure and cryptorchidism in Canadian populations. *Environmental Health Perspectives.* 2017 May; 125(5): 057004.

242. Poon S, Koren G, Carnevale A, Aleksa K, Ling J, et al. Association of in utero exposure to polybrominated diphenyl ethers with the risk of hypospadias. *JAMA Pediatrics.* 2018;172(9):851-856.

243. Whorton MD. Male occupational reproductive hazards. *Western Journal of Medicine.* 1982;137(6):521–524.

244. Chan SL. Male infertility: Diagnosis and treatment. *Canadian Family Physician.* 1988;34:1735–1738.

245. Products—Vital Statistics Rapid Release—Natality Quarterly Provisional Estimates. 2018; https://www.cdc.gov/nchs/nvss/vsrr/natality-dashboard.htm.

246. Bichell RE. Average age of first-time moms keeps climbing

in the U.S. NPR. 2018; https://www.npr.org/sections/health-shots/2016/01/14/462816458/average-age-of-first-time-moms-keeps-climbing-in-the-u-s.

247. Chandra A, Copen CE, Stephen EH. Infertility and impaired fecundity in the United States, 1982–2010: Data from the National Survey of Family Growth. *National Health Statistics Report.* 2013(67):1–18, 1 p following 19.

248. Slama R, Hansen OK, Ducot B, et al. Estimation of the frequency of involuntary infertility on a nation-wide basis. *Human Reproduction.* 2012;27(5):1489–1498.

249. Sunderam S, Kissin DM, Crawford SB, et al. Assisted reproductive technology surveillance — United States, 2014. *Morbidity and Mortality Weekly Report Surveillance Summaries.* 2017;66(6):1–24.

250. Schwartz A. People aren't having babies in Denmark so they made this hilariously provocative ad. *Business Insider.* October 2, 2015; http://www.businessinsider.com/do-it-for-denmark-ad-campaign-to-encourage-pregnancy-2015-10.

251. Levine H, Jorgensen N, Martino-Andrade A, et al.

252. Jurewicz J, Radwan M, Sobala W, et al. Human urinary phthalate metabolites level and main semen parameters, sperm chromatin structure, sperm aneuploidy and reproductive hormones. *Reproductive Toxicology (Elmsford, NY).* 2013;42:232–241.

253. Wirth JJ, Rossano MG, Potter R, et al. A pilot study associating urinary concentrations of phthalate metabolites and semen quality. *Systems Biology in Reproductive Medicine.* 2008;54(3):143–154.

254. Hauser R, Meeker JD, Duty S, et al. Altered semen quality in relation to urinary concentrations of phthalate monoester and oxidative metabolites. *Epidemiology.* 2006;17(6):682.

255. Joensen UN, Frederiksen H, Jensen MB, et al. Phthalate excretion pattern and testicular function: A study of 881 healthy Danish men. *Environmental Health Perspectives.* 2012;120(10):1397–1403.

256. Jonsson BA, Richthoff J, Rylander L, et al. Urinary phthalate metabolites and biomarkers of reproductive function in young men. *Epidemiology.* 2005;16(4):487–493.

257. Ramezani Tehrani F, Noroozzadeh M, Zahediasl S, et al. The

time of prenatal androgen exposure affects development of polycystic ovary syndrome-like phenotype in adulthood in female rats. *International Journal of Endocrinology and Metabolism.* 2014;12(2):e16502.

258. Abbott DH, Barnett DK, Bruns CM, et al. Androgen excess fetal programming of female reproduction: A developmental aetiology for polycystic ovary syndrome? *Human Reproduction Update.* 2018;11(4):357–374.

259. Wu Y, Zhong G, Chen S, et al. Polycystic ovary syndrome is associated with anogenital distance, a marker of prenatal androgen exposure. *Human Reproduction.* 2017;32(4):937–943.

260. Hotchkiss AK, Lambright CS, Ostby JS, et al. Prenatal testosterone exposure permanently masculinizes anogenital distance, nipple development, and reproductive tract morphology in female Sprague-Dawley rats. *Toxicology and Science.* 2007;96(2):335–345.

261. Mendiola J, Stahlhut RW, Jørgensen N, et al. Shorter anogenital distance predicts poorer semen quality in young men in Rochester, New York. *Environmental Health Perspectives.* 2011;119(7):958–963.

262. Fisher JS, Macpherson S, Marchetti N, et al. Human 'testicular dysgenesis syndrome': A possible model using in-utero exposure of the rat to dibutyl phthalate. *Human Reproduction.* 2003;18(7):1383–1394.

263. Li N, Chen X, Zhou X, et al. The mechanism underlying dibutyl phthalate induced shortened anogenital distance and hypospadias in rats. *Journal of Pediatric Surgery.* 2015;50(12):2078–2083.

264. Martino-Andrade AJ, Liu F, Sathyanarayana S, et al. Timing of prenatal phthalate exposure in relation to genital endpoints in male newborns. *Andrology.* 2016;4(4):585–593.

265. Bornehag CG, Carlstedt F, Jonsson BA, et al. Prenatal phthalate exposures and anogenital distance in Swedish boys. *Environmental Health Perspectives.* 2015;123(1):101–107.

266. Buck Louis GM, Sundaram R, Sweeney AM, et al. Urinary bisphenol A, phthalates, and couple fecundity: The Longitudinal Investigation of Fertility and the Environment (LIFE) Study. *Fertility and Sterility.* 2014;101(5):1359–1366.

267. Attina TM, Hauser R, Sathyanarayana S, et al.

268. Konkel L. Reproductive headache? Investigating acetaminophen as a potential endocrine disruptor. *Environmental Health Perspectives*. 2018;126(3):032001.

269. Kristensen DM, Hass U, Lesne L, et al. Intrauterine exposure to mild analgesics is a risk factor for development of male reproductive disorders in human and rat. *Human Reproduction*. 2011;26(1):235–244.

270. Snijder CA, Kortenkamp A, Steegers EA, et al. Intrauterine exposure to mild analgesics during pregnancy and the occurrence of cryptorchidism and hypospadia in the offspring: The generation R study. *Human Reproduction*. 2012;27(4):1191–1201.

271. Kristensen DM, Mazaud-Guittot S, Gaudriault P, et al. Analgesic use—prevalence, biomonitoring and endocrine and reproductive effects. *Nature Reviews Endocrinology*. 2016;12(7):381–393.

272. Nigg JT, Lewis K, Edinger T, et al. Meta-analysis of attention-deficit/hyperactivity disorder or attention-deficit/hyperactivity disorder symptoms, restriction diet, and synthetic food color additives. *Journal of the American Academy of Child and Adolescent Psychiatry*. 2012;51(1):86–97.e88.

273. Kristensen DM, Desdoits-Lethimonier C, Mackey AL, et al. Ibuprofen alters human testicular physiology to produce a state of compensated hypogonadism. *Proceedings of the National Academy of Sciences of the United States of America*. 2018;115(4):e715–e724.

274. Barbuscia A, Mills MC. Cognitive development in children up to age 11 years born after ART—A longitudinal cohort study. *Human Reproduction*. 2017;32(7):1482–1488.

275. Guo XY, Liu XM, Jin L, et al. Cardiovascular and metabolic profiles of offspring conceived by assisted reproductive technologies: A systematic review and meta-analysis. *Fertility and Sterility*. 2017;107(3):622–631.e625.

276. Ferguson KK, McElrath TF, Meeker JD. Environmental phthalate exposure and preterm birth. *JAMA Pediatrics*. 2014;168(1):61–67.

277. Meeker JD, Ferguson KK.

278. Feldman HA, Goldstein I, Hatzichristou DG, et al. Impotence

and its medical and psychosocial correlates: Results of the Massachusetts Male Aging Study. *Journal of Urology*. 1994;151(1):54–61.

279. Espir ML, Hall JW, Shirreffs JG, et al. Impotence in farm workers using toxic chemicals. *BMJ*. 1970;1(5693):423–425.

280. Polsky JY, Aronson KJ, Heaton JP, et al. Pesticides and polychlorinated biphenyls as potential risk factors for erectile dysfunction. *Journal of Andrology*. 2007;28(1):28–37.

281. Li D, Zhou Z, Qing D, et al. Occupational exposure to bisphenol-A (BPA) and the risk of self-reported male sexual dysfunction. *Human Reproduction*. 2010;25(2):519–527.

282. Li DK, Zhou Z, Miao M, et al. Relationship between urine bisphenol-A level and declining male sexual function. *Journal of Andrology*. 2010;31(5):500–506.

283. Igharo OG, Anetor JI, Osibanjo O, et al. Endocrine disrupting metals lead to alteration in the gonadal hormone levels in Nigerian e-waste workers. *Universa Medicina*. 2018;37(1).

284. Bergman A, Heindel JJ, Jobling S, et al.

285. Gore AC, Chappell VA, Fenton SE, et al.

6. THE CHEMICAL VULNERABILITY OF GIRLS AND WOMEN

286. Buck Louis G, Cooney MA, Peterson CM. Ovarian dysgenesis syndrome. *Journal of Developmental Origins of Health and Disease*. 2011;2(01):25–35.

287. Buck Louis G, Cooney M. Effects of environmental contaminants on ovarian function and fertility. In González-Bulnes A, ed. *Novel concepts in ovarian endocrinology*. Kerala, India: Transworld Research Network, 2007.

288. Gravholt CH, Andersen NH, Conway GS, et al. Clinical practice guidelines for the care of girls and women with Turner syndrome: Proceedings from the 2016 Cincinnati International Turner Syndrome Meeting. *European Journal of Endocrinology*. 2017;177(3):G1–G70.

289. Buck Louis G, Cooney M. Ovarian dysgenesis syndrome. *Journal of Developmental Origins of Health and Disease*. 2011;2(01):25–35.

290. Rogers PA, Adamson GD, Al-Jefout M, et al. Research priorities for endometriosis. *Reproductive Sciences.* 2017;24(2):202–226.

291. Adamson GD, Kennedy S, Hummelshoj L. Creating solutions in endometriosis: Global collaboration through the World Endometriosis Research Foundation. *Journal of Endometriosis and Pelvic Pain Disorders.* 2018;2(1):3–6.

292. Hunt PA, Sathyanarayana S, Fowler PA, Trasande L.

293. Kawwass JF, Monsour M, Crawford S, et al. Trends and outcomes for donor oocyte cycles in the United States, 2000–2010. *JAMA.* 2013;310(22):2426–2434.

294. Davies C, Godwin J, Gray R, et al. Relevance of breast cancer hormone receptors and other factors to the efficacy of adjuvant tamoxifen: Patient-level meta-analysis of randomised trials. *Lancet.* 2011;378(9793):771–784.

295. Gamble J. Puberty: Early starters. *Nature.* 2017;550(7674):S10–S11.

296. Bellis MA, Downing J, Ashton JR. Adults at 12? Trends in puberty and their public health consequences. *Journal of Epidemiology and Community Health.* 2006;60(11):910–911.

297. Marshall WA, Tanner JM. Variations in pattern of pubertal changes in boys. *Archives of Disease in Childhood.* 1970;45(239):13–23.

298. Marshall WA, Tanner JM. Variations in pattern of pubertal changes in girls. *Archives of Disease in Childhood.* 1969 Jun; 44(235):291–303.

299. Herman-Giddens ME, Slora EJ, Wasserman RC, et al. Secondary sexual characteristics and menses in young girls seen in office practice: A study from the Pediatric Research in Office Settings network. *Pediatrics.* 1997;99(4):505–512.

300. Harley KG, Rauch SA, Chevrier J, et al. Association of prenatal and childhood PBDE exposure with timing of puberty in boys and girls. *Environment International.* 2017;100:132–138.

301. Wolff MS, Pajak A, Pinney SM, et al. Associations of urinary phthalate and phenol biomarkers with menarche in a multiethnic cohort of young girls. *Reproductive Toxicology.* 2017;67:56–64.

302. Katz TA, Yang Q, Treviño LS, Walker CL, Al-Hendy A. Endocrine-disrupting chemicals and uterine fibroids. *Fertility and Sterility.* 2016 Sep 15; 106(4): 967-977.

303. Baird DD, Newbold R. Prenatal diethylstilbestrol (DES) exposure is associated with uterine leiomyoma development. *Reproductive Toxicology*. 2005 May–Jun; 20(1): 81–4.

304. Fowler PA, Childs AJ, Courant F, et al. In utero exposure to cigarette smoke dysregulates human fetal ovarian developmental signalling. *Human Reproduction (Oxford, England)*. 2014;29(7):1471–1489.

305. Peretz J, Vrooman L, Ricke WA, et al. Bisphenol A and reproductive health: Update of experimental and human evidence, 2007–2013. *Environmental Health Perspectives*. 2014;122(8):775–786.

306. Pat Hunt and Sheela Sathyanarayana were leading one group of experts to examine effects of EDCs on female reproduction that focused on fibroids and endometriosis. They were also joined by Paul Fowler, a feisty Scottish biologist at the University of Aberdeen; Ruthann Rudel, an epidemiologist based at the Silent Spring Institute, a Massachusetts-based institute focused on environmental factors and women's health; and Barbara Cohn, a California-based epidemiologist. Joining them as well were Ana Soto, a developmental biologist at Tufts University, and Frederic Bois, a toxicologist based at the French National Institute for Environment and Risk.

307. Fei X, Chung H, Taylor HS. Methoxychlor disrupts uterine Hoxa10 gene expression. *Endocrinology*. 2005;146(8):3445–3451.

308. Takayama S, Sieber SM, Dalgard DW, et al. Effects of long-term oral administration of DDT on nonhuman primates. *Journal of Cancer Research and Clinical Oncology*. 1999;125(3–4):219–225.

309. Bredhult C, Backlin BM, Bignert A, et al. Study of the relation between the incidence of uterine leiomyomas and the concentrations of PCB and DDT in Baltic gray seals. *Reproductive Toxicology (Elmsford, NY)*. 2008;25(2):247–255.

310. Hunt PA, Sathyanarayana S, Fowler PA, Trasande L.

311. Trabert B, Chen Z, Kannan K, et al. Persistent organic pollutants (POPs) and fibroids: Results from the ENDO study. *Journal of Exposure Science & Environmental Epidemiology*. 2015;25(3):278–285.

312. Hunt PA, Sathyanarayana S, Fowler PA, Trasande L.

313. Kassotis CD, Tillitt DE, Davis JW, et al. Estrogen and androgen

receptor activities of hydraulic fracturing chemicals and surface and ground water in a drilling-dense region. *Endocrinology.* 2014;155(3):897–907.

314. Kassotis CD, Klemp KC, Vu DC, et al. Endocrine-disrupting activity of hydraulic fracturing chemicals and adverse health outcomes after prenatal exposure in male mice. *Endocrinology.* 2015;156(12):4458–4473.

315. Kassotis CD, Bromfield JJ, Klemp KC, et al. Adverse reproductive and developmental health outcomes following prenatal exposure to a hydraulic fracturing chemical mixture in female C57Bl/6 mice. *Endocrinology.* 2016;157(9):3469–3481.

316. Balise VD, Meng CX, Cornelius-Green JN, et al. Systematic review of the association between oil and natural gas extraction processes and human reproduction. *Fertility and Sterility.* 2016;106(4):795–819.

317. Casey JA, Savitz DA, Rasmussen SG, et al. Unconventional natural gas development and birth outcomes in Pennsylvania, USA. *Epidemiology.* 2016;27(2):163–172.

318. McKenzie LM, Guo R, Witter RZ, et al. Birth outcomes and maternal residential proximity to natural gas development in rural Colorado. *Environmental Health Perspectives.* 2014;122(4): 412–417.

319. Currie J, Greenstone M, Meckel K. Hydraulic fracturing and infant health: New evidence from Pennsylvania. *Science Advances.* 2017;3(12):e1603021.

320. Bouwman H, van den Berg H, Kylin H. DDT and malaria prevention: Addressing the paradox. *Environmental Health Perspectives.* 2011;119(6):744–747.

321. The recruitment methods could have introduced biases that could otherwise explain results found in less-than-optimal studies. Some studies relied on self-report of endometriosis, which can be a big problem because the gold standard is to see the endometriosis surgically. Especially for outcomes that are either yes or no, being imprecise in how you measure exposure can actually decrease chances of finding important effects statistically.

322. Buck Louis GM, Peterson CM, Chen Z, et al. Bisphenol A and phthalates and endometriosis: The Endometriosis: Natural History, Diagnosis and Outcomes Study. *Fertility and Sterility.* 2013;100(1):162–169.e161–162.

323. Liu J, Wang W, Zhu J, et al. Di(2-ethylhexyl) phthalate (DEHP) influences follicular development in mice between the weaning period and maturity by interfering with ovarian development factors and microRNAs. *Environmental Toxicology.* 2018;33(5):535–544.

324. Hunt PA, Sathyanarayana S, Fowler PA, Trasande L.

325. Attina TM, Hauser R, Sathyanarayana S, et al.

326. Swaen GMH, Otter R. Letter to the editor: Phthalates and endometriosis. *Journal of Clinical Endocrinology and Metabolism.* 2016;101(11):L108–l109.

327. Hunt PA, Sathyanarayana S, Fowler PA, et al. Response to the letter by G. M. H. Swaen and R. Otter. *Journal of Clinical Endocrinology and Metabolism.* 2016;101(11):L110–L111.

328. American Cancer Society. Economic impact of cancer. 2018; https://www.cancer.org/cancer/cancer-basics/economic-impact-of-cancer.html.

329. Birrer N, Chinchilla C, Del Carmen M, et al. Is hormone replacement therapy safe in women with a BRCA mutation?: A systematic review of the contemporary literature. *American Journal of Clinical Oncology.* 2018;41(3):313–315.

330. McNeil M. Menopausal hormone therapy: Understanding long-term risks and benefits. *JAMA.* 2017;318(10):911–913.

331. Grossman DC, Curry SJ, Owens DK, et al. Hormone therapy for the primary prevention of chronic conditions in postmenopausal women: US Preventive Services Task Force Recommendation Statement. *JAMA.* 2017;318(22):2224–2233.

332. Kauff ND, Satagopan JM, Robson ME, et al. Risk-reducing salpingo-oophorectomy in women with a BRCA1 or BRCA2 mutation. *New England Journal of Medicine.* 2002;346(21):1609–1615.

333. Cohn BA, Wolff MS, Cirillo PM, et al. DDT and breast cancer in young women: New data on the significance of age at exposure. *Environmental Health Perspectives.* 2007;115(10):1406–1414.

334. USEPA. Pesticides Industry Sales and Usage. 2008-12 Market Estimates. https://www.epa.gov/sites/production/files/2017-01/documents/pesticides-industry-sales-usage-2016_0.pdf.

335. Tyrone Hayes and Penelope Jagessar Chaffer give the science a thorough and clear treatment in a 2010 TedWomen talk: https://www.ted.com/talks/tyrone_hayes_penelope_jagessar_chaffer_the_toxic_baby.

336. Ibarluzea J, Fernández M, Santa-Marina L, et al. Breast cancer risk and the combined effect of environmental estrogens. *Cancer Causes Control.* 2004;15(6):591–600.

337. Pastor-Barriuso R, Fernandez MF, Castano-Vinyals G, et al. Total effective xenoestrogen burden in serum samples and risk for breast cancer in a population-based multicase-control study in Spain. *Environmental Health Perspectives.* 2016;124(10):1575–1582.

338. Trasande L, Massey RI, DiGangi J, et al. How developing nations can protect children from hazardous chemical exposures while sustaining economic growth. *Health Affairs.* 2011;30(12):2400–2409.

7. REAL STEPS THAT MAKE A DIFFERENCE

339. Whyatt RM, Rauh V, Barr DB, et al.

340. Lu C, Toepel K, Irish R, et al.

341. Bradman A, Quiros-Alcala L, Castorina R, et al.

342. Environmental Working Group. EWG's 2018 Shopper's Guide to Pesticides in Produce. 2018; https://www.ewg.org/foodnews/.

343. Bradman A, Quiros-Alcala L, Castorina R, et al.

344. Agriculture Marketing Service, US Department of Agriculture. National Bioengineered Food Disclosure Standard. Available at https://www.federalregister.gov/documents/2018/05/04/2018-09389/national-bioengineered-food-disclosure-standard (Accessed August 12 2018).

345. Rudel RA, Gray JM, Engel CL, et al.

346. Environmental Working Group. Skin Deep Cosmetics Database 2018; https://www.ewg.org/skindeep/#.WtzE_dTwa6I.

347. Harley KG, Kogut K, Madrigal DS, et al.

348. Morgan MK, Jones PA, Calafat AM, et al.

349. Rudel RA, Gray JM, Engel CL, et al.

350. Carwile JL, Ye X, Zhou X, et al. Canned soup consumption and urinary bisphenol A: A randomized crossover trial. *JAMA.* 2011;306(20):2218–2220.

351. Schecter A, Malik N, Haffner D, et al.

352. Trasande L. Further limiting bisphenol A in food uses could provide health and economic benefits.

353. Kuruto-Niwa R, Nozawa R, Miyakoshi T, et al.

354. Chen MY, Ike M, Fujita M.

355. Yoshihara Si, Mizutare T, Makishima M, et al.

356. Okuda K, Fukuuchi T, Takiguchi M, et al.

357. Audebert M, Dolo L, Perdu E, et al.

358. Danzl E, Sei K, Soda S, et al.

359. Ike M, Chen MY, Danzl E, et al.

360. WHO. Sustaining the elimination of iodine deficiency disorders (IDD). 2007; http://www.who.int/nmh/iodine/en/.

361. Rogan WJ, Paulson JA, Baum C, et al. Iodine deficiency, pollutant chemicals, and the thyroid: New information on an old problem. *Pediatrics.* 2014;133(6):1163–1166.

362. Sathyanarayana S, Alcedo G, Saelens BE, et al. Unexpected results in a randomized dietary trial to reduce phthalate and bisphenol A exposures. *Journal of Exposure Science & Environmental Epidemiology.* 2013;23(4):378–384.

363. Galloway TS, Baglin N, Lee BP, et al. An engaged research study to assess the effect of a 'real-world' dietary intervention on urinary bisphenol A (BPA) levels in teenagers. *BMJ Open.* 2018;8(2):e018742.

364. Hodson R. Precision medicine. *Nature.* 2016;537(7619):S49.

8. YOUR VOICE MATTERS:
HOW YOU CAN PARTICIPATE IN A VIRTUOUS CIRCLE

365. There are more details in references 73–78 for the reader interested in details. The papers published in peer-reviewed journals describe it in even more detail, but Monte Carlo modeling is a widely used method to account for multiple independent

scenarios in which there might be such costs that are uncertain.

366. Attina TM, Hauser R, Sathyanarayana S, et al.

367. Eliperin J, Dennis B. White House eyes plan to cut EPA staff by one-fifth, eliminating key programs. 2017; https://www.wash ingtonpost.com/news/energy-environment/wp/2017/03/01/white-house-proposes-cutting-epa-staff-by-one-fifth-eliminat ing-key-programs/ (Accessed June 29, 2017).

368. Forman J, Silverstein J. Organic foods: Health and environmental advantages and disadvantages. *Pediatrics*. 2012;130(5):e1406–1415.

369. Zota AR, Calafat AM, Woodruff TJ. Temporal trends in phthalate exposures: Findings from the National Health and Nutrition Examination Survey, 2001–2010. *Environmental Health Perspectives*. 2014;122(3):235–241.

370. Centers for Disease Control and Prevention. National report on human exposure to environmental chemicals. https://www .cdc.gov/exposurereport/.

371. Zota AR, Phillips CA, Mitro SD. Recent fast food consumption and bisphenol A and phthalates exposures among the U.S. population in NHANES, 2003–2010. *Environmental Health Perspectives*. 2016;124(10):1521–1528.

372. Serrano SE, Braun J, Trasande L, et al.

373. Public Health Institute. Green cleaning in schools: A guide for advocates. http://www.phi.org/uploads/application/files/khc qbtguo1fuyi5w10wortxqfpnrwrsode32y7sbqsocfbouyo.pdf.

374. NRDC Greening Advisor. Safer chemicals: Pesticides & fertilizers.2018;http://nba.greensports.org/safer-chemicals/pesticides -fertilizers/.

375. Badenhausen K. NBA team values 2018: Every club now worth at least $1 billion. *Forbes*. 2018; https://www.forbes.com/sites/kurtbadenhausen/2018/02/07/nba-team-values-2018-every-club-now-worth-at-least-1-billion/#5e8a76957155.

376. @ceh4health. Major producers eliminating flame retardant chemicals as major buyers are demanding flame retardant-free furniture—Center for Environmental Health. 2018.

377. Wadhwa V. The big lesson from Amazon and Whole Foods: Disruptive competition comes out of nowhere. *MarketWatch.* https://www.marketwatch.com/story/the-big-lesson-from-amazon-and-whole-foods-disruptive-competition-comes-out-of-nowhere-2017-06-19.

378. Painter KL. Americans are eating more organic food than ever, survey finds. *Minneapolis Star Tribune.* 2018; http://www.startribune.com/americans-are-eating-more-organic-food-than-ever-survey-finds/424061513/

379. Center for Food Safety and Applied Nutrition (FDA). Labeling & nutrition. https://www.fda.gov/Food/GuidanceRegulation/GuidanceDocumentsRegulatoryInformation/LabelingNutrition/ucm456090.htm (Accessed June 12, 2018).

380. Ball J. What is 'natural' food anyway? *Irish Times.* 2018; https://www.irishtimes.com/life-and-style/food-and-drink/what-is-natural-food-anyway-1.3154859 (Accessed June 12, 2018).

381. Center for Food Safety and Applied Nutrition (FDA). Labeling & nutrition.

382. USDA. Meat and poultry labeling terms. 2018; https://www.fsis.usda.gov/wps/portal/fsis/topics/food-safety-education/get-answers/food-safety-fact-sheets/food-labeling/meat-and-poultry-labeling-terms/meat-and-poultry-labeling-terms.

383. USDA. Questions and answers. USDA shell egg grading service. Agricultural Marketing Service. 10/15; https://www.ams.usda.gov/publications/qa-shell-eggs.

384. FDA. Everything added to food in the United States (EAFUS). 2018; https://www.accessdata.fda.gov/scripts/fcn/fcnNavigation.cfm?rpt=eafusListing.

385. Kraft Heinz Company. Kraft Heinz expands environmental commitments to include sustainable packaging and carbon reduction. https://news.kraftheinzcompany.com/press-release/corporate/kraft-heinz-expands-environmental-commitments-include-sustainable-packaging- (Accessed September 29 2018).

386. Schenk M, Popp SM, Neale AV, et al. Environmental medicine content in medical school curricula. *Academic Medicine.* 1996;71(5):499–501.

387. Roberts JR, Gitterman BA. Pediatric environmental health education: A survey of US pediatric residency programs. *Ambulatory Pediatrics.* 2003;3(1):57–59.

388. Bergman Å, Heindel JJ, Jobling S, Kidd KA, Zoeller RT, eds.

389. WHO. NCDs | Web-based consultation (May 10–16, 2018). http://www.who.int/ncds/governance/high-level-commission/web-based-consultation-may2018/en/ (Accessed June 12, 2018).

390. Waters E, de Silva-Sanigorski A, Hall BJ, et al. Interventions for preventing obesity in children. *Cochrane Database of Systematic Reviews.* 2011(12):Cd001871.

391. Wild CP. Complementing the genome with an "exposome": The outstanding challenge of environmental exposure measurement in molecular epidemiology. *Cancer Epidemiology, Biomarkers & Prevention.* 2005;14(8):1847–1850.

INDEX